Contents

the NEW APPALACHIA: IDEAS that WORK

Think of this small book as a collection of action snapshots of a region on the move. Sometimes the focus is on the land itself, such as "brownfield" sites in western Pennsylvania or small farms in southern New York. Often it's on infrastructure, like a replacement for an aging bridge in Tennessee or links between roads, rail, and a waterway in northeastern Mississippi. Frequently the camera zooms in on sophisticated technology, such as laptops used by schoolchildren in Georgia or satellite-assisted surveys in western Maryland.

Always, always, we see people in action—working together to build something. What they're building may be a structure, like a water line along a rocky ridge in western Virginia; or a strategy, like a Kentucky program designed to produce homegrown doctors for rural Appalachia. Sometimes they're investing in projects whose payoffs may not materialize for a generation, as with a West Virginia youth leadership camp or an Alabama county's efforts to encourage its talented high school graduates to stay in the area.

One way or another, all these stories are about capacity building—acts of faith in the future of Appalachia. They're also about collaboration within communities, across the Appalachian Region, and with partners in the larger world outside Appalachia. In that respect, they're evidence of how Appalachia has changed during the 37 years since the President's Appalachian Regional Commission called Appalachia "a region apart" from the rest of America. But they're also examples of a continuing commitment to the vision that made change possible.

In 1965, economically speaking, Appalachia's eggs were in a very few baskets, each vulnerable to market shocks. The Region depended heavily on the extraction of natural resources and on agriculture. In the southern states, manufacturing meant mostly low-wage textile mills; in the northern Rust Belt, it meant heavy industry in aging plants employing fewer and fewer workers. From 1950 to 1960, a decade when national employment grew 15 percent, Appalachian employment actually declined. One in three Appalachians lived in poverty, a rate 50 percent higher than the national average. The Region's narrow mountain roads choked off the growth of commerce and industry and constricted people's access to jobs, schools, and services. They were used by trucks hauling coal and timber to railheads, and, all too often, by some of the Region's most talented young people moving to places far away.

In creating the Appalachian Regional Commission (ARC), the Congress mandated a partnership between the federal government and Appalachian states, to be reinforced by a unique governance structure. All ARC decisions are approved by two chairpersons—a presidential appointee confirmed by the Senate and a governor selected by his or her fellow Appalachian governors.

ARC spearheaded an assault on isolation. Its first priority became the design and routing of a network of modern, four-lane roads known collectively as the Appalachian Development Highway System (ADHS). The system's explicit purpose was economic and human development. Corridor routes were chosen with an eye to opening up isolated areas, not adding capacity to places where traffic volumes were already high. Today about 2,330 miles of the ADHS are open to traffic. A 1998 study of the economic development impact of completed segments of 12 ADHS corridors (about 1,400 miles) showed that they have made travel easier and safer, are creating thousands of new jobs, and are projected to generate over a billion dollars in economic development benefits.

At the same time, ARC and its state partners invested in the Region's capacity for growth. Then, as now, that often included infrastructure: water and sewer projects, community facilities, and industrial parks. It also included direct investments in people: adult education, health services, and leadership training.

The changes in Appalachia have been dramatic. Since 1969, employment in the Region has grown by more than 50 percent. The Region's poverty rate has been cut in half; by 1990 it differed from that of the rest of the nation by only two percentage points. Thanks to better water and better medical care, infant mortality has fallen by more than two-thirds. The percentage of Appalachian adults with at least a high school education has more than doubled; and for young adults (ages 18 to 24) this percentage is slightly higher than the U.S. rate. A 1995 study showed that Appalachian counties have grown significantly faster than a comparison group of their socioeconomic "twins" outside the Region.

Nevertheless, nearly 700 miles of the ADHS are still incomplete, and hundreds of communities still lack access to clean water. Of the Region's 406 counties, 114 remain economically distressed. Most of these lie at the Region's center, which still suffers from isolation and the poverty isolation fosters.

In 1996 ARC reinvented itself. That is, it adopted a strategic plan that reaffirmed its two historic priorities—an economic and human development program and a developmental highway program—and rethought how to achieve them. The plan established goals under five headings: 1) education and work-force training, 2) physical infrastructure, 3) civic capacity and leadership, 4) dynamic local economies, and 5) health care.

This book shows some of the ways in which states and local communities are fleshing out the states' strategies for reaching the Region's goals in economic and human development. Many of the projects this book describes—such as a telemedicine program and an initiative to market Appalachian products in Latin America, Europe, and Asia—would have been impossible to imagine in 1965. Yet the original vision remains as it was: to end isolation and to help Appalachian communities develop the capacity to renew themselves. The federal-state-community collaboration this vision brought into being is not only intact, but is also often broadened by private-sector initiatives. The projects described in this book show that collaboration in action and prove that the Region's capacity for self-renewal is as strong as ever.

GOAL ONE

Education and Workforce Training

Appalachian residents will have the skills and knowledge necessary to compete in the world economy in the twenty-first century.

ALABAMA

GETTING CHILDREN READY TO LEARN: ALABAMA SCHOOL READINESS PROGRAM

Many parents in Appalachian Alabama do not have access to quality preschool programs. As a result, their children often begin school unprepared to take full advantage of their classroom experience. To address this critical issue, the state's Office of School Readiness in the Department of Children's Affairs is providing more academic preschool programs in 13 Appalachian counties, with help from an array of partners including the Appalachian Regional Commission, state and local government agencies, and private businesses. This effort involves identifying potential grantees among nonprofit child-care providers, and then building the capacity of these providers to prepare children for school. The project also offers technical assistance, training, and curricular support to selected centers. Parent training workshops, following national PTA guidelines, focus on improving parenting skills and encouraging parents to become more involved in their children's school experience.

CONTACT:
Betsy Taff
Director
Office of School Readiness
Alabama Department of Children's Affairs
RSA Tower, Suite 1670
Montgomery, Alabama 36130-2755
(334) 353-1095
Email: tsmith@dca.state.al.us

GEORGIA

MAXIMIZING SUCCESS IN THE TWENTY-FIRST CENTURY: NORTH GEORGIA TECHNICAL COLLEGE ENTREPRENEURIAL EDUCATION PROGRAM

Small business startups have increased over the past few years, and this trend will continue as new opportunities arise in the high-tech and service industries. Recognizing the importance of helping start new businesses in northern Georgia, the North Georgia Technical College has developed an entrepreneurial education program. Participants gain a better understanding of the business world and what it takes to increase their chances for success once they start out on their own. The entrepreneurial education program offers a number of courses for academic credit. These include 16 courses available over the Internet through the Georgia Virtual Technical Institute. In addition, the program operates a Small Business Resource Center open to students and aspiring entrepreneurs.

CONTACT:
Fran Chastain
Entrepreneurial Education Program Director
North Georgia Technical College
P.O. Box 65
Clarkesville, Georgia 30523
(706) 754-7810
Email: fchas@ngtcollege.org

GOAL ONE PROJECTS

INVESTING IN HUMAN CAPITAL: EARLY CHILDHOOD INITIATIVE IN APPALACHIAN GEORGIA

Five counties in Appalachian Georgia are taking a multidimensional, multigenerational approach to improving the lives of families with young children. Faced with high rates of poverty, teen pregnancy, illiteracy, and unemployment, these counties are working to improve services in five key areas: universal contact at birth; intensive home visitation; developmental child care; parenting education; and adult education and job training. During a nine-month period beginning in July 2000, more than 400 families with newborns were contacted; more than 300 families received home visits; more than 90 children received developmental child care; more than 80 parents enrolled in GED classes; and more than 30 child-care providers received training. All services are voluntary, and families are encouraged to participate as full partners in determining their strengths, needs, and future goals.

CONTACT:
Carol C. Wilson
State Director
Early Childhood Initiative
409 Bear Paw Trail
Blue Ridge, Georgia 30513
(706) 632-7342
Email: carolcwilson@tds.net

BREATHING NEW LIFE INTO OLD PRACTICES: COMPUTER-AIDED DESIGN FOR THE GRANITE INDUSTRY

Granite is the major industry and primary source of employment in Elbert County. Stonecutting, sandblasting, etching, and polishing are traditional skills that have been practiced until recently with traditional tools and line-of-sight measurements. Over the last decade, however, computer-aided design methods have entered the granite industry. Many draftspeople trained in traditional methods are retiring, and a new generation of employees will soon be needed. The Elbert County Comprehensive High School set up a new work-training program so that students can learn state-of-the-art computer-aided design methods from skilled industry artisans. Local industry is working with the school, which purchased 26 specialized computer stations and created a new etching class as well. In the first year, 40 students learned computer-aided design, and more than ten mastered specialized granite-etching skills.

CONTACT:
Nancy Bessinger
Tech Career Director
Elbert County Comprehensive High School
600 Abernathy Circle
Elberton, Georgia 30635
(706) 283-3680
Email: nbessing@elbert.k12.ga.us

CONNECTING SCHOOLS AND COMMUNITY: THE CHATTOOGANET PROJECT

In Chattooga County, Georgia, local businesses, government agencies, and community groups have joined forces in support of ChattoogaNet, an Internet service provider (ISP) run by local students. The program not only teaches participating students the mechanics of operating an Internet server, but also provides free Internet access to all segments of the community, including businesses, local government agencies, and residents. The students receive intensive technology training and fully participate in helping manage the ISP's day-to-day operation. As part of their training, students learn Web page development, a talent reflected in the Chattooga community Web page. In addition to improving technical skills, the program also teaches students the importance of learning partnership skills, sharing resources, and contributing to the community.

CONTACTS:
Rick Clifton or Diane Bryant
The ChattoogaNet Project
Chattooga County Schools
33 Middle Road
Summerville, Georgia 30747
(706) 895-3340
Email: rclifton@chattooga.k12.ga.us
    dbryant@chattooga.k12.ga.us

WORKING TOGETHER FOR OUR CHILDREN: THE PARENTS, CHILDREN, AND TEACHERS FORCE 3 PROGRAM

One proven way to improve school performance is to encourage more parents to play a greater role in their children's education, in cooperation with their children's teachers. At three targeted schools in Chattooga County, Georgia, officials have introduced a new program designed to improve parental accountability in children's education, while also strengthening parent and teacher communications. Called the Parents, Children, and Teachers (PCT) Force 3 Program, the initiative focuses on parents of children considered at risk of poor school performance. To date, more than 300 parents have participated in workshops and activities offered at schools and community centers. Some activities focused only on parents, with child care provided, while other activities included parents and children. By the end of the school year, teachers had made over 250 separate home or work-site visits to parents of at-risk children.

CONTACT:
Tonia Reynolds
PCT Force 3 Program Parent Coordinator
Chattooga County Schools
P.O. Box 30
Summerville, Georgia 30747
(706) 857-3490
Email: treynolds@chattooga.k12.ga.us

GOAL ONE PROJECTS

EXPANDING THE CLASSROOM: TOWNS COUNTY MIDDLE SCHOOL LAPTOP COMPUTER PROJECT

With only a few computers in the library, students in Towns County Middle School had little opportunity to develop computer skills or use the Internet to research projects. In 1998, in the first year of a pilot program, Towns County gave every middle school student a specially designed laptop computer. The school also offered training for teachers, students, and parents, and provided access to the Internet from school or home through a school-based network. Computers are now as ubiquitous as books in the classroom. Students, their parents, and teachers have found their lives enriched in unexpected ways. Since every student has access to online research, teachers can assign more interesting and demanding projects. Parents are also learning computer skills and communicating with their children's teachers through email. Students and parents both report that the children are spending less time watching television and more time doing homework. A number of parents have been inspired to resume their own education. The successful project has led to similar investments in middle schools in West Virginia and North Carolina. In May 2001, *Time* magazine ranked Towns County Middle School among the most innovative middle schools in the country.

CONTACT:
Stephen H. Smith
Principal
Towns County Middle School
1400 U.S. Highway 76, East
Hiawassee, Georgia 30546
(706) 896-4131
Email: SSmith@mail.towns.kl2.ga.us

KENTUCKY

HELPING STUDENTS AIM HIGH: CHALLENGER LEARNING CENTER IN HAZARD, KENTUCKY

Launched in the spring of 1999, this hands-on education project offers thousands of middle school students the opportunity to participate in an eight-week science and math curriculum, aligned with state and national standards, that prepares them for a simulated space mission. The program teaches basic math and science skills and motivates students to pursue careers in these fields. The first of its kind in rural America, the Hazard center is part of a network of over 30 Challenger Learning Centers nationwide begun by a not-for-profit education organization founded in 1986 by the surviving family members of the seven Challenger astronauts. To date, more than 18,000 students have participated in the space simulation program.

CONTACT:
Tom Cravens
Director
Challenger Learning Center of Kentucky
601 Main Street
Hazard, Kentucky 41701
(800) 334-2793
(606) 436-5721
Email: tom.cravens@kctcs.net

MARYLAND

PROVIDING SATELLITE TECHNOLOGY AT LOCAL COLLEGES TO IMPROVE LAND-USE PLANNING: GLOBAL POSITIONING SYSTEM

Many people know that satellite-based global positioning system (GPS) technology is revolutionizing navigation. Fewer may realize that it is also changing how surveyors, mapping professionals, and land-use planners carry out their work. GPS technology can be especially helpful in mountainous areas such as Appalachia because a line of sight between surveying stations is not required. To provide the data necessary for GPS surveying and geographic information systems (GIS) mapping, officials in Allegany, Garrett, and Washington Counties, in western Maryland, have established survey-quality base stations at three community colleges, not only to train students but also to make these services available free to private companies. The base stations, one of which is part of a national GPS grid established by the U.S. Geological Survey, provide GPS data through the Internet to private surveyors, land planners, and other users. Several different GPS receivers have been purchased. Both mapping and surveying machines are available for student and professional use. Funds generated by renting field equipment and providing reduced-rate training classes to the industry help support the GPS project.

CONTACT:
Steve Resh
Coordinator of Forestry Programs
Allegany College of Maryland
12401 Willowbrook Road
Cumberland, Maryland 21502
(301) 784-5307
Email: sresh@ac.cc.md.us

HELPING LOCAL GOVERNMENT PLANNERS: GEOGRAPHIC INFORMATION SYSTEMS

Computerized geographic information systems (GIS) technology is becoming an integral part of land planning and environmental management for county and municipal governments in western Maryland. This is the direct result of a state initiative making current computerized geographical data available to local governments. A key part of the initiative is a specialized GIS laboratory at Frostburg State University in western Maryland. Frostburg's Department of Geography provides low-cost GIS development, staff training, and student internships. The Tri-County Council for Western Maryland hopes to offer GIS support to economic development departments within the county in the near future.

CONTACT:
Guy Winterberg
Grantsperson
Tri-County Council for Western Maryland
111 South George Street
Cumberland, Maryland 21502
(301) 777-2158
Email: gwinterberg@tccwmd.org

GOAL ONE PROJECTS

MISSISSIPPI

EDUCATING ENTREPRENEURS: MISSISSIPPI STATE UNIVERSITY EXTENSION SERVICES' VIRTUAL ENTREPRENEURIAL EDUCATION AND TRAINING PROGRAM

Providing effective training to help Appalachians start and maintain their own businesses is critical for rural communities that seek to grow and diversify their economies. The Mississippi State University Extension Service provided these services to seven counties through its Virtual Entrepreneurial Education and Training program. The project initially brought together a variety of resources to help launch new businesses and provide entrepreneurial training to over 200 youths and adults in the seven-county target area. Youths had access to extensive training and the opportunity to participate in cross-age teaching and learning. Adults were trained in "Entrepreneurship 101" and "E-Biz" curricula with extensive e-commerce educational support, including access to laptop computers, Web development software, interactive video, and Internet marketing tools. Youths were trained through the establishment of the Entrepreneurship Corps (E-Corps). E-Corps participants were taught extensive technology and entrepreneurial skills that enabled some of them to start new business ventures. The training ended with several of the entrepreneurs selling their products on the Internet through retail Web sites.

CONTACTS:
Beth Duncan
Small Business Specialist
Mississippi State University Extension Service
Box 9642
Mississippi State, Mississippi 39762
(662) 325-2160
Email: bethd@ext.msstate.edu

Linda Mitchell
Youth Technology and Special Programs Coordinator
Mississippi State University Extension Service
P.O. Box 2297
5338 Cliff Gookin Boulevard
Tupelo, Mississippi 38803
(601) 841-9000
Email: lindam@ext.msstate.edu

DEVELOPING A BETTER-EDUCATED WORKFORCE: ITAWAMBA COMMUNITY COLLEGE ADVANCED EDUCATION CENTER

According to recent research, more than one-third of the value-added jobs in Mississippi are located within a 50-mile radius of Tupelo. Recognizing its critical role in improving workforce skills in the area, Itawamba Community College has expanded its capacity to meet the training needs of more than 400 employers and 500 workers in northeastern Mississippi by opening the Advanced Education Center (AEC) at its Tupelo campus. With support from local leaders and funding from the Appalachian Regional Commission, the center has been equipped with an array of modern resources, including a fluid power lab, basic and advanced programmable logic controller labs, and four computer labs. Working closely with the University of Mississippi and the Mississippi University for Women, the AEC anticipates that by developing a better-educated workforce, Itawamba County and its neighboring counties will improve their local economies.

CONTACT:
Charles Chrestman
Vice President for Instruction
Itawamba Community College
602 West Hill Street
Fulton, Mississippi 38843
(662) 862-8050
Email: cvchrestman@icc.cc.ms.us

NEW YORK

SHOWING WHY MATH AND SCIENCE ARE IMPORTANT: ALLEGANY COUNTY TECHNOLOGY CURRICULUM IMPLEMENTATION PROJECT

Responding to the growing demand for students proficient in mathematics, science, and technology, the Cuba-Rushford Central School District in Allegany County, New York, has sponsored a number of student-centered projects that demonstrate the practical applications of scientific inquiry and research. In 1998, a hatchery was built for the artificial propagation and rearing of young fish. The student research center, which includes a dock and observation deck, facilitates outdoor and indoor research using the latest technology. Eighth-grade students engage in problem solving, data collection and analysis, field research, statistical analysis, interdisciplinary connections, and computer use. When the program is fully operational, all 800 students at the Cuba-Rushford Central School are expected to participate.

CONTACT:
Scott Jordan
Science Teacher
Cuba-Rushford Central School
5476 Route 305
Cuba, New York 14727
(716) 968-2650
Email: sjordan@crcs.wnyric.org

GOAL ONE PROJECTS

LINKING NETWORKS TO IMPROVE EDUCATION: LEATHERSTOCKING TELECOMMUNICATIONS CONSORTIUM

What is now the Leatherstocking Telecommunications Consortium began as several distinct networking efforts linking classrooms in multiple school districts. It evolved into a sophisticated telecommunications system providing Internet access, distance learning, telemedicine services, and mobile teleconferencing for local governments and businesses. Over the last decade, quickly evolving technology has presented challenges and opportunities for regional telecommunications planners. Students continue to benefit from "distant" classes while new technology expands the network's capabilities. The consortium also has helped local governments set up Internet sites providing citizen access to government information. An associated health care telecommunications network provides links to hospitals and outreach clinics that support nursing education and provide telemedicine assistance to public school nurses. The consortium is now in the process of expanding the network to nine counties. Plans include delivering health education to schools, storing training on a video server, and linking college classrooms to a video network.

CONTACT:
Stan France
Director of Central Data Processing
Schoharie County
P.O. Box 541
Schoharie, New York 12157
(518) 295-8465
Email: stan@co.schoharie.ny.us

LINKING STUDENTS AND BUSINESS: WHITESVILLE SCHOOL-BUSINESS PARTNERSHIP

Once a solid farming community, the rural hamlet of Whitesville in southeastern Allegany County is experiencing economic difficulty and can offer only limited job opportunities to its high school graduates. To stimulate stronger ties between students and local businesses, the Whitesville Central School District has created a special school-business partnership that links art and technology classes with the needs of small businesses in the area. Equipped with computers and special software, purchased with support from the Appalachian Regional Commission, the school's print shop provides students an opportunity to learn the basics of commercial printing, graphics, and advertising while producing brochures and pamphlets for area businesses. The students not only gain valuable work experience but also develop personal ties to local businesses.

CONTACTS:
Charles Cutler
Superintendent
Whitesville Central School
692 Main Street
Whitesville, New York 14897
(607) 356-3301 ext. 221

Douglas Van Skiver
School Counselor
Whitesville Central School
692 Main Street
Whitesville, New York 14897
(607) 356-3301 ext. 234
Email: doug_vanskiver@whitesville.wnyric.org

TRAINING MACHINISTS FOR THE COMPUTER AGE: COMPUTER NUMERICAL CONTROL MACHINE TOOL LABORATORY

In Allegany and Cattaraugus Counties, the New York Department of Labor forecasts a growing demand for machinists with computer numerical control (CNC) training. The demand will be especially acute with the retirement of current machinists, 50 percent of whom are age 50 or older. Alfred State College, at the request of local companies and the Allegany County Employment and Training Center, has launched an advanced machine tool certificate and machine tool technology degree program. Specialized instruction will include CNC machine programming, CNC lathe operation, CNC milling machine operation, and computer-aided design and computer-aided manufacturing processes. The county employment and training center expects to be able to place at least 20 trained graduates a year. In addition, 30 to 45 students a year will be enrolled in machinist coursework through contract courses with regional industries and training agencies.

CONTACT:
Craig Clark
Dean of Vocational Technology
Alfred State College
School of Vocational Technology
2530 South Brooklyn Avenue
Wellsville, New York 14895
(607) 587-3101
Email: clarkcr@alfredstate.edu

OVERCOMING BARRIERS TO LEARNING: CYBER ADULT LEARNING PROJECT

Close to a quarter of adults in Delaware and Otsego Counties have no high school diploma, a much higher percentage than in surrounding areas. These adults face many barriers to education. They do not have access to public transportation; most have limited financial resources; and those with children have limited time. To overcome these barriers and increase the number of adults capable of passing GED testing, the Oneonta Community Education Center in Otsego County, an adult literacy partnership, has developed New York State's first adult education Internet site. The Cyber Adult Learning Project allows those over the age of 16 to study and qualify online for the GED. During the first week, a plan is developed for individual students based on prior experience and existing skills. Computers are assigned to the adult learners. Students gain access to the education resources necessary for completing their GED and also develop critical computer skills.

CONTACT:
Cathy Jeanette
Program Coordinator
Oneonta Community Education Center
10 Market Street
Oneonta, New York 13820
(607) 433-3645
Email: jeanetcm@oneonta.edu

GOAL ONE PROJECTS

NORTH CAROLINA

CREATING NEW CAMPUSES ON THE INFORMATION HIGHWAY: ALLEGHANY CYBER SITE

Students in Alleghany, Wilkes, and Ashe Counties in North Carolina have not always had direct access to the resources of the state's university system. North Carolina is now leveraging those university resources by making information and classes available to students throughout the state. Alleghany High School is one of seven cyber campuses to be equipped with multimedia interactive computer equipment linked to the North Carolina School of Science and Mathematics and the Internet through the high-speed, fiber-optic North Carolina Information Highway. The campus is also linked to Wilkes Community College and provides distance-learning classes for high school students, teachers, and the general community. On-site computer training is also provided. The cyber campus, which is open evenings and weekends for general public and business users, served approximately 6,886 students and over 1,430 adults (including businesspersons, government workers, and the general public) in 1998, its first year in business.

CONTACT:
Phil Trew
Senior Regional Planner
Region D Council of Governments
P.O. Box 1820
Boone, North Carolina 28607
(828) 265-5434
Email: ptrew@regiond.org

BRINGING COMPUTERS TO RURAL DAY-CARE CENTERS: REGION I EARLY CHILDHOOD DEVELOPMENT REGIONAL NETWORK

Though fast-changing information technology is transforming government and private-sector management practices, many nonprofits have limited equipment and expertise. Many nonprofits are unable to provide computer literacy opportunities to the people they serve, those who are least likely to have computers available elsewhere. These weaknesses became obvious in 1996 when local community and government leaders struggled to develop telecommunications plans in the first phase of North Carolina's Connect NC initiative. The leaders recommended creation of the Early Childhood Development Regional Network that now provides networked computers in the 12 day-care centers overseen by the Northwest Child Development Council. The network is transforming and streamlining management of the centers: office software has been standardized; long-distance costs cut; accounts-payable processing time trimmed; and inventory and supply management centralized. Children now learn with the help of computers. The staff are computer literate, as are many parents who take advantage of after-hours access to computers and the Internet.

CONTACT:
Gary Steeley
Director of Information Services
Northwest Piedmont Council of Governments
400 West Fourth Street, Suite 400
Winston-Salem, North Carolina 27101
(336) 761-2111
Email: gsteeley@nwpcog.dst.nc.us

PREPARING CHILDREN FOR SCHOOL: REGION A PARTNERSHIP FOR CHILDREN (SMART START)

When North Carolina first funded Smart Start in 1993, many of the state's children were physically and socially unprepared for school. Close to 20 percent of the children lived in poverty, many were not receiving immunizations, the mortality rate was high, and almost 10,000 families waited to enroll for subsidized child-care assistance. Rather than mandate a solution to these many problems, Smart Start required counties to establish local community boards to create and run local Smart Start programs. This decision presented a challenge to all counties, especially those in Appalachian North Carolina. Nevertheless, with support from the Appalachian Regional Commission, most of the 29 Appalachian counties had programs in place by 1996. Boards were able to analyze, plan, and create partnerships to win future funding for their programs. The Region A Partnership for Children was a pioneer in the statewide initiative and continues to help meet a wide range of needs for services including child-care training, assistance, and referral; health and dental services; parenting training; and coordinated family services.

CONTACT:
June Smith
Executive Director
Region A Partnership for Children
116 Jackson Street
Sylva, North Carolina 28779
(828) 586-0661
Email: jtsmith9@gte.net

MAKING COLLEGE ACCESSIBLE: NEW CENTURY SCHOLARS

Many rural high school students are qualified to attend college but do not go for a variety of reasons. Some are reluctant to become the first in their families to attend college; others worry about the expense and feel pressure to enter the full-time workforce immediately after high school. New Century Scholars, a cooperative effort of the business community, the public schools, and local colleges, is providing a new educational path for students in southwestern North Carolina. Starting in the seventh grade, students selected to participate agree to meet education and conduct standards that will entitle them to receive a college scholarship. Initiated in 1995, the program at present includes 600 New Century Scholars, who attend area schools. The first of these enrollees is attending Southwestern Community College this fall. The number of students in the program (which usually adds 120 to 140 new enrollees each year) depends upon the amount of scholarship money available to support the effort. Funds are raised in the community through individual pledges. The program includes intervention with the selected students, as well as parent involvement. Students who perform satisfactorily in high school attend the community college. If they complete an associate's degree, they spend their final two years at a state university. The incentive to stay in school has reduced dropout rates. In 1999, the New Century Scholars program leveraged a $2.4 million grant from the U.S. Department of Education to expand the program.

CONTACT:
Connie Haire
Vice President for Student and Institutional Development
Southwestern Community College
447 College Drive
Sylva, North Carolina 28779
(828) 586-4091 ext. 227
Email: connie@southwest.cc.nc.us

GOAL ONE PROJECTS

OHIO

TURNING CAMPUSES INTO YOUTH COMMUNITY CENTERS: KIDS ON CAMPUS COMMUNITY PARTNERSHIP

Kids on Campus was initiated in 1996 to meet the needs of low-income elementary school children in Athens County. The Kids on Campus Community Partnership has hosted between 300 and 450 children on the campuses of Ohio University and Hocking College each summer. The curriculum features exploratory math, science, and computer labs; literacy labs featuring individualized tutoring; recreation and wellness activities; fine arts; and a new engineering and robotics component entitled the Cranium Challenge. Students receive daily meals and health screenings that include hearing and vision testing. The program provides leadership training for parents, college and high school students, and teachers. Prior to each summer session all staff and volunteers receive an intensive week of training in leadership, conflict resolution, child development, behavior management, and multidisciplinary curriculum development. Kids on Campus offers diverse educational and recreational opportunities for the underserved children and youths of Athens County.

CONTACT:
Ann Teske
Project Director
Kids on Campus
Ohio University
Grosvenor Hall, Room 19
Athens, Ohio 45701
(614) 593-9335
Email: teske@ohio.edu

EXPANDING COMPUTER LEARNING TO MEET SPECIFIC INDUSTRY NEEDS: JEFFERSON COMMUNITY COLLEGE ENGINEERING COMPUTER PROJECT

Based on successful recruiting efforts in Steubenville, Ohio, several regional manufacturing and design firms concluded that Jefferson Community College in Steubenville would be an excellent source for highly motivated and competent employees. Hoping to hire more employees, the Steubenville firms recommended that the college update its computer-aided design (CAD) laboratory and establish an engineering computation laboratory. In its first year, the new CAD facility directly benefited over 220 design students. More than 1,100 students were served by the engineering computation laboratory, through new courses in computer science and significant enhancements in the college's advanced mathematics, science, and engineering curricula.

CONTACT:
Laura Meeks
President
Jefferson Community College
4000 Sunset Boulevard
Steubenville, Ohio 43952
(740) 264-5591
Email: lmeeks@jefferson.cc.oh.us

KEEPING COMPUTER SKILLS UP TO DATE: SWISS HILLS CAREER CENTER COMPUTER LAB

Recognizing that local companies in Monroe, Belmont, and Noble Counties need employees experienced in using current business software, the Swiss Hills Career Center has worked to meet the demand. In addition to purchasing 150 interactive devices and a $250,000 rapid prototyping device, the center has updated and expanded curricula and included a certified Cisco training program. Swiss Hills has been a central force in bringing close to 1,000 new and predominantly high-tech jobs to Monroe County in recent years. Student scores on the Ohio Proficiency Test also have improved, and postsecondary enrollment has increased from 15 to more than 60 percent.

CONTACT:
Terry A. Wallace
Director
Swiss Hills Career Center
46601 SR 78
Woodsfield, Ohio 43793
(740) 472-0722
Email: docta96@ovnet.com

HELPING WOMEN IMPROVE THEIR JOB SKILLS: UNIVERSITY OF CINCINNATI CLERMONT WORKFORCE DEVELOPMENT

The majority of students at the University of Cincinnati's Clermont College campus are women. A typical student is in her late twenties or thirties, attends school part-time, works part-time, and cares for her children. Many of the women seek training in order to get better jobs. An analysis of area companies found that computer training was a primary need of employers and employees, as well as a priority for those in the state's school-to-work and welfare-to-work programs, in which Clermont College participates. In response to these needs, the school launched an $11 million workforce development program that includes an expanded staff and various new resources, such as a learning facility featuring 43,000 square feet of classrooms and laboratories. The Appalachian Regional Commission provided support to equip two new computer labs and expand the staff serving the learning center, which can now accommodate 1,000 students. Based on past experience, officials are optimistic about these new efforts. Historically, close to half of the participants in skills-training programs at Clermont continue toward a bachelor's degree. Most of the remaining participants enter or return to the workforce with new skills and higher earning potential.

CONTACT:
Elaine Mueninghoff
Associate Dean of Academic Services
University of Cincinnati Clermont College
4200 Clermont College Drive
Batavia, Ohio 45103-1785
(513) 732-5212
Email: muenine@email.uc.edu

GOAL ONE PROJECTS

MAKING UNIVERSITY KNOWLEDGE WORK FOR COMMUNITIES: INSTITUTE FOR LOCAL GOVERNMENT ADMINISTRATION AND RURAL DEVELOPMENT AT OHIO UNIVERSITY

Since 1981, the nationally recognized Institute for Local Government Administration and Rural Development (ILGARD) has expanded the capacity of local governments and nonprofit agencies in the 29 Appalachian Ohio counties to serve their communities. Using Ohio University knowledge and other resources, ILGARD provides small communities the same access to applied research and technical assistance as larger, wealthier communities. Assistance includes developing and implementing geographic information systems (GIS) technology; training; collecting data; conducting survey research, strategic planning, and program evaluations; and facilitating public meetings. ILGARD staff, along with Ohio University faculty and students, is currently working on 70 projects. Projects include the highly regarded restoration of the 116-square-mile Monday Creek Watershed, a considerable environmental project for Ohio. ILGARD has also helped establish a priority list of problems, coordinate volunteer projects, and use its GIS capabilities to create interactive maps of the watershed. ILGARD studies have highlighted numerous issues that have led to important regional initiatives, such as the Ohio Appalachian Center for Higher Education and the Appalachian Partnership for Welfare Reform.

CONTACT:
Mark Weinberg
Director of Voinovich Center
Institute for Local Government Administration and Rural Development
143 Technology and Enterprise Building
20 East Circle Drive
Athens, Ohio 45701
(740) 593-4388
Email: weinberg@ilgard.ohiou.edu

MOTIVATING STUDENTS TO ATTEND COLLEGE: OHIO APPALACHIAN CENTER FOR HIGHER EDUCATION

Recognizing that residents of Ohio's 29 Appalachian Ohio counties attended college at a rate far below the state and national rates, the state government and a consortium of ten public colleges and universities formed the Ohio Appalachian Center for Higher Education (OACHE) to motivate more students to attend college. The program focuses on helping both students and teachers overcome the psychological barriers that often prevent students at rural schools from considering college. Since the program's inception in 1993, grantee schools have increased their college-attendance rates by an average of 16 percent. Among these grantees, one of the poorest schools in Ohio, Southern Local High School in Meigs County, achieved a 92 percent college-attendance rate in 2001. In 1998, the Community Colleges of Appalachia, in partnership with Bluefield State College, opened the North Central Appalachian Center for Higher Education at Bluefield State College to sponsor OACHE-like programs with partner schools in West Virginia and Appalachian Maryland. Following the success of that pilot center, the West Virginia Access Center for Higher Education (WVACHE) was established. In 2000, the Appalachian Regional Commission funded a program to replicate the OACHE/WVACHE model in other Appalachian states. In May 2001, the Public Employees Roundtable, a national coalition of organizations representing public employees and retirees, awarded OACHE its Public Service Excellence Award for outstanding state program in 2001.

CONTACT:
Wayne F. White
Executive Director
Ohio Appalachian Center for Higher Education
Shawnee State University
The Commons
Portsmouth, Ohio 45662
(740) 355-2299
(866) 466-2243
Email: WWhite@Shawnee.edu

PENNSYLVANIA

HELPING UNIVERSITIES HELP LOCAL BUSINESSES: PENNSYLVANIA ALLIANCE OF HIGHER EDUCATION FOR RESEARCH AND TECHNOLOGY

Higher education institutions contain many resources valuable to industry, but private companies often have problems identifying and using the expertise that universities offer. In response to this situation, the Pennsylvania Alliance of Higher Education for Research and Technology (PART) created the first higher education research and technology database accessible via the Internet. The PART Web site catalogues research and technology facilities, faculty, and equipment available to businesses and industry under a variety of partnering agreements at 22 northeastern Pennsylvania colleges and universities. A variety of partners support the effort. These include participating schools, Ben Franklin Technology Partners of Northeastern Pennsylvania, the Northeastern Pennsylvania Alliance, the Northeastern Pennsylvania Industrial Resource Center, and the Technology Council of Northeastern Pennsylvania.

CONTACT:
Penny Cannella
Regional Coordinator
Pennsylvania Alliance of Higher Education for Research and Technology
1151 Oak Street
Pittston, Pennsylvania 18640
(570) 655-5581
Email: cannella@nepa-alliance.org

GOAL ONE PROJECTS

IMPROVING SKILLS AND PERFORMANCE: SOUTHERN ALLEGHENIES WORKFORCE CONSORTIA PROJECT

Realizing that skilled employees are a company's most valuable resource, the Southern Alleghenies Workforce Consortia project is helping more than 80 employers in six Pennsylvania counties improve job-training programs. Working in 12 different job sectors including manufacturing, health care, and technology, the project has helped form industry-specific groups whose members meet regularly to determine training needs and identify common areas of greatest need. Areas already identified include safety enforcement, communications, computer and information technology, customer service, and supervisory skills. The project so far has received funding for training in one skill area and is seeking additional funding to provide training in other areas.

CONTACT:
Terri Campbell
Workforce Development Services Manager
Southern Alleghenies Planning and Development Commission
541 58th Street
Altoona, Pennsylvania 16602
(814) 949-6710
Email: Campbell@sapdc.org

TRAINING DISABLED ADULTS FOR NEW JOBS: PARTNERS IN PROGRESS

Partners in Progress (PIP) has provided vocational training to disabled adults in Tioga County since 1997, and many of their clients serve regional businesses by performing light assembly duties for manufacturing production. The organization has been so successful that it has outgrown its facility. A partnership of customer companies and local development organizations is helping raise money to purchase and renovate a former supermarket to house expanded services and production. Renovations will eliminate a blighted building and allow PIP to serve customers and clients from a centralized location. Thirty-seven jobs will be saved, and PIP expects to double employment in the next three years. The group hopes to provide additional work-training and employment opportunities through new outreach in Tioga County and other counties, thereby providing additional services to regional manufacturers as well as meaningful employment for disabled adults.

CONTACT:
Jill Koski
Economic Development Program Manager
Northern Tier Regional Planning and Development Commission
312 Main Street
Towanda, Pennsylvania 18848
(570) 265-9103
Email: koski@northerntier.org

USING NEW TECHNOLOGIES TO ADVANCE EDUCATION AND IMPROVE ECONOMIC DEVELOPMENT: THE SCHOOLHOUSE PROJECT IN GREENE COUNTY

A partnership between a local high school in Greene County, Pennsylvania, and a technology company provides science students the opportunity to use an advanced scanning electron microscope (SEM) during the day and conduct materials-characterization analysis for paying customers in the evening. In addition, this unique project is helping encourage small business development at an associated technology incubator. In the Schoolhouse Project, the West Greene School District has developed an innovative science-to-work program based on a new laboratory facility and an SEM.

The RJ Lee Group has provided equipment, funding, expertise, and training. A high-speed Internet connection links the corporate and school sites, allowing tutoring, demonstrations, and access to other schools and businesses. The company employs high school and college students to assist with educational and commercial project work. The professional materials-characterization work program for students in the Greene County region is part of a comprehensive workforce development plan in this distressed county. The program and the lab's capabilities have helped encourage high-technology development in the area. A planned business incubator at EverGreene Technology Park will be home to small companies taking advantage of the laboratory, technical assistance, shared services, and availability of trained employees.

CONTACT:
Brian Jackson
Principal
West Greene High School
Waynesburg, Pennsylvania 15370
(724) 499-5191
Email: jacksonb@west-greene.k12.pa.us

TENNESSEE

BUILDING A CENTER FOR NEW OPPORTUNITIES: MARION COUNTY ADULT EDUCATION AND SKILLS TRAINING CENTER

In the early 1990s, one-third of the adult population in Marion County did not have a high school diploma, and the dropout rate for high school seniors was almost 7 percent. In 1991 the county built the Marion County Adult Education and Skills Training Center. Over 500 people have received their GED or adult high school diploma, and up to 300 people a day now use the facility, which houses the adult education program, vocational rehabilitation services, and satellite offices for the Private Industry Council, the Department of Human Services, and Chattanooga State Community College. Through basic education classes, college-level courses, and direct employment training, many residents of Marion County and surrounding counties are better equipped to enter the job market or improve their employment.

CONTACT:
Howell Moss
County Executive
Marion County
P.O. Box 789
Jasper, Tennessee 37347
(423) 942-2552
Email: mcexec@bellsouth.net

GOAL ONE PROJECTS

VIRGINIA

MEETING DIVERSE NEEDS: SOUTHWEST VIRGINIA HIGHER EDUCATION CENTER'S REGIONAL TRAINING AND CONFERENCE CENTER

Realizing they shared a common need for quality space to conduct an array of activities, local business, government, and education leaders in southwestern Virginia joined together to build the Regional Training and Conference Center at the Southwest Virginia Higher Education Center. The state-of-the-art complex includes classroom facilities with interactive telecommunications technology and a multipurpose conference hall that accommodates 1,500 people. Using distance-learning technologies and in-class instruction, partner schools have expanded their executive and adult training as well as courses for degree-seeking students. Numerous local companies have conducted employee training; professional continuing education seminars have been provided; public meetings have been held; and business and trade shows have been hosted. The Regional Training and Conference Center satisfies a cluster of needs that were unlikely to be have been met individually.

CONTACT:
Rachel Fowlkes
Executive Director
Southwest Virginia Higher Education Center
P.O. Box 1987
One Partnership Circle
Abingdon, Virginia 24212
(540) 469-4005
Email: rfowlkes@swcenter.edu

WEST VIRGINIA

PROVIDING EMPLOYMENT FOR WORKERS WITH DISABILITIES: THE HANCOCK COUNTY SHELTERED WORKSHOP

The Hancock County Sheltered Workshop (HCSW) is helping local businesses while providing work for individuals with disabilities. The workshop is a nonprofit organization founded in 1952 by parents who wanted to provide academic and social activities for their children with mental or developmental disabilities. Today it is one of the leading community rehabilitation programs in the state, meeting the diverse needs of 86 individuals with disabilities. Seeking meaningful new work opportunities for its clients, HCSW determined there was local demand for commercial laundry services. With local and state support, and funding in 2000 from the Appalachian Regional Commission and the U.S. Department of Agriculture, the workshop constructed a state-of-the-art commercial laundry facility with the capacity to process up to five million pounds of linen annually. By spring 2001, the facility was operating at a rate of one million pounds per year and already providing services for four major customers, including a hospital, two nursing homes, and a private business. More importantly, the laundry provided work for 15 full-time employees, including eight individuals with disabilities, and 50 part-time employees, including 35 with developmental disabilities. HCSW expects the facility to reach full capacity by July 2002, providing services for an expanding network of local businesses and institutions and offering additional jobs to people with disabilities.

CONTACT:
Fred Hendershot
Executive Director
Hancock County Sheltered Workshop
1100 Pennsylvania Avenue
Weirton, West Virginia 26062
(304) 748-2370
Email: fhenders@1st.net

EXPANDING CURRICULA AND INCREASING TECHNOLOGY USE: RIVERSIDE HIGH SCHOOL

Students participating in the fine arts cluster at the new Riverside High School in eastern Kanawha County can now produce their own daily television news programs. They have learned to use cameras, lighting, computer graphics, teleprompters, and video-editing equipment purchased with help from the Appalachian Regional Commission and Kanawha County. The program was made possible following a consolidation effort in 1999, when the county replaced two older institutions and combined grades 9 through 12 at the new high school. The new configuration has allowed greater flexibility and opportunity for curriculum development. The studio and technology facilities now provide training and exposure to professional fields never before available to area students. The new school's technology lab, used by an estimated 700 students a year, also provides community education and professional workshops for area adults. The Upper Kanawha Valley Chamber of Commerce, the Upper Kanawha Valley Mayors' Association, and the Upper Kanawha Valley Economic Development Corporation were involved in planning the new school, and continue to support its innovative programs, which benefit students and encourage regional economic development.

CONTACT:
Richard Clendenin
Principal
Riverside High School
#1 Warrior Way
Belle, West Virginia 20515
(304) 348-1996
Email: rclendon@rhs.kana.k12.wv.us

GOAL ONE PROJECTS

MIXING FUN WITH YOUTH LEADERSHIP TRAINING: CAMP HORSESHOE

The students who spend time at Camp Horseshoe in St. George, West Virginia, each summer go home motivated to make their communities better. Youths participating in a special leadership and civic development initiative run by the Ohio–West Virginia High School YMCA (HI-Y) Leadership Center must work at least 25 hours in community service upon their return home. Students tutor, work on a Rails-to-Trails project, help senior citizens, and raise money to enable other students to attend Camp Horseshoe. Civic training for West Virginia students culminates with an annual Youth in Government program in Charleston. Economic and business leadership is also developed at Camp Horseshoe, which since 1978 has conducted the Free Enterprise Conference. In 2000, 188 students from West Virginia participated in three one-week leadership development summits that focused on improving entrepreneurial skills and encouraging students to start their own businesses. Each of these youngsters was involved with at least one youth group of 25 persons or more in his or her home area, dramatically extending the impact of the skills and training these students received. Program organizers estimate that these students will engage in at least 11,000 hours of volunteer service in their communities, using the grant-writing training and other skills developed in the leadership program.

CONTACT:
David King
Executive Director
Ohio–West Virginia YMCA
Route 2, Box 138
St. George, West Virginia 26287
(304) 478-2481
Email: hiymail@hiyleads.org

INCREASING WORK-BASED SKILLS: CLAY COUNTY SCHOOLS

High schools students in Clay County had little opportunity to develop work-related skills, either in school or in local businesses. School officials decided to employ a work-site coordinator to identify potential work sites, train work-site mentors, implement computer-simulated workplaces, and develop school-based enterprises. An initial grant from the Appalachian Regional Commission helped school officials hire a coordinator and purchase computer hardware and workplace-simulation software. These initial efforts helped lead to the award of a $1.3 million, four-year U.S. Department of Labor grant to support School-to-Work opportunities. In four years, the number of high school seniors participating in work-related experiences at school or community sites has increased from 7 to 128. Of these, close to 100 seniors have received credit in classes where computer-assisted workplace simulations composed a significant part of the curriculum.

CONTACT:
Jeff Krauklis
Assistant Superintendent
Clay County Schools
242 Church Street
Clay, West Virginia 25043
(304) 587-4266
Email: jkrauk@hotmail.com

PROVIDING CHILD CARE FOR ECONOMIC DEVELOPMENT: CHILDREN FIRST CHILD DEVELOPMENT CENTER

"My children need to come first." This was the recurring refrain from women employees of several Jefferson County businesses, brought together in focus groups to discuss their needs as working parents. The economy was growing, labor in this eastern panhandle county was becoming scarce, and area businesses needed to attract and retain good employees. For working mothers, child care was a big concern. As many as 3,100 children needed child care, but there were only 671 licensed slots in the area. The Jefferson County Development Authority helped local employers, other local agencies, and citizens respond by creating the Children First Child Development Center. This independent, not-for-profit, tax-exempt organization is dedicated to providing working parents and their children affordable, quality childhood care and education. The effort is expected to bring new employees into the workforce, reduce absenteeism, and increase productivity for participating companies. Local monies, supplemented with a grant from the Appalachian Regional Commission, funded a 6,000-square-foot building on donated land in the Burr Industrial Park. The facility can serve up to 100 children age six weeks and older. It will offer continuing care through elementary school age, primarily for children of company employees but also for the greater community, as space is available. This public-private partnership benefits the community, working mothers, employers, and most of all, children.

CONTACT:
Jane Peters
Executive Director
Jefferson County Development Authority
P.O. Box 237
Charles Town, West Virginia 25412
(304) 728-3255
Email: jcda@intrepid.net

GOAL TWO

Physical Infrastructure

Appalachian residents will have the physical infrastructure necessary for self-sustaining economic development and improved quality of life.

ALABAMA

WORKING TOGETHER FOR MUTUAL BENEFIT: RAINSVILLE INDUSTRIAL PARK INFRASTRUCTURE

With the closure of five textile plants in De Kalb County during a recent two-year period, local community leaders recognized the need to attract other industry to the Rainsville Industrial Park. By improving the park's water supply and sewer capacity, community leaders were able to develop a $10 million facility to make interior and exterior plastic parts for a new Honda assembly plant in nearby Lincoln. Officials expect at least 120 new jobs to be created at the new Rainsville Technology, Inc., plastics manufacturing site, a subsidiary of Moriroku Company Limited of Japan. The new plant would not have been possible without the cooperation of two municipal governments—the town of Section, which owned the water system, and the city of Rainsville, which ran the sewer system. Although some municipal governments can be territorial about their water and sewer services, these two municipalities agreed to work together to develop the new plant. As a result, both communities are now enjoying the benefits of that cooperation.

CONTACT:
Kim Erwin
Consultant
Morton and Associates
200 East McKinney Avenue
Albertville, Alabama 35950
(256) 878-5222
Email: mortonl@strnt.net

MAKING ECONOMIC GROWTH POSSIBLE: LINCOLN WASTEWATER TREATMENT PLANT

Honda of America's decision to locate a new plant near Lincoln, Alabama, was great news for the community. Initially employing 1,500 workers, the plant is expected to create as many as 5,000 additional jobs as Honda suppliers open facilities in the area. An economic impact study performed by Auburn University's Center for Government and Public Affairs estimates that these 1,500 jobs will generate state and local taxes of $4 million and state sales taxes of $2.5 million. A new wastewater treatment plant, adjacent to the Honda site, is needed to provide adequate treatment capacity for this new development. It will also serve approximately 2,300 existing residences and 75 existing commercial customers, in addition to handling projected growth in Lincoln's population. The new plant will treat up to two million gallons of wastewater a day with ultraviolet radiation, providing better treatment than the old system and improving water quality in the Coosa River. The city of Lincoln, which will own and operate the wastewater treatment plant, will fund construction in conjuction with state and federal agencies and the Appalachian Regional Commission.

CONTACT:
Donna Fathke
Principal Planner
East Alabama Regional Planning and Development Commission
P.O. Box 2186
Anniston, Alabama 36202
(256) 237-6741
Email: dfathke@adss.state.al.us

GOAL TWO PROJECTS

WIRING FOR BUSINESS, EDUCATION, AND HEALTH: BIG SANDY REGIONAL TELECOMMUNICATIONS CENTER

Determined that the information highway not bypass Pike County and the Big Sandy area of eastern Kentucky, community leaders created the Big Sandy Regional Telecommunications Center. Operated by the nonprofit Big Sandy Telecommuting Services, Inc., the center currently provides a variety of services, including hands-on computer, network, and Internet training and certification; coordination of remote teaching and teleconferencing; and business support services. Partners in the telecommunications center include the Pikeville College School of Osteopathic Medicine, whose Telemedicine Services and Learning Center at the site will soon serve faculty, students, and the public. In addition, to support new business development, the new facility is taking on the additional role of business incubator, providing office space and access to shared personnel and equipment to several new, emerging enterprises.

CONTACT:
Sandy Runyon
Executive Director
Big Sandy Area Development District
100 Resource Drive
Prestonsburg, Kentucky 41653
(606) 886-2374
Email: sandyr@bigsandy.adds.state.ky.us

BUILDING A NEW LOCAL ECONOMY: ADVANCED TECHNOLOGY CENTER AND TECHNICAL INNOVATION CENTER

In the mid 1980s, after Fairchild Industries closed its aircraft manufacturing operations, people in Washington County began working to develop and attract new high-tech industries. The Advanced Technology Center was opened at Hagerstown Community College in 1990, and the Technical Innovation Center, a major $2 million addition, was completed in 1994. By 1995, six high-tech enterprises in such diverse fields as chemicals, electronics, and computer software were using the new facilities. By 1997, the Advanced Technology Center had become fully self-supporting. From its rental and service revenues the facility has installed broadband Internet networking connectivity throughout the building. At the Technical Innovation Center, entrepreneurs can take an idea through the stages of computer-aided design, development, test marketing, and production. In addition, local businesses have access to advanced technical resources as well as state-supported economic development agencies. Ultimately, local officials say, the center will lead to a better-trained, more adaptable workforce. It has already directly assisted over 40 firms that now employ 170 people.

CONTACT:
P. Chris Marschner
Manager
Technical Innovation Center
11400 Robinswood Drive
Hagerstown, Maryland 21742-6590
(301) 790-2800 ext. 479
Email: marschnerc@hcc.cc.md.us

PRESERVING THE PAST FOR A MORE PROSPEROUS FUTURE: CANAL PLACE HERITAGE AREA

For almost 75 years, the C&O Canal played a major role in defining the economy of Cumberland, Maryland, and the surrounding region. Today, as a result of the efforts of many partners, the canal is once again a major focal point for economic development in Allegany County. Recognizing Cumberland's role in American history as the western terminus of the Chesapeake and Ohio Canal, community leaders created Canal Place, the state's first heritage area initiative. By forming strong partnerships with federal, state, and local agencies, including the C&O Canal National Historical Park and the Appalachian Regional Commission, the Canal Place Preservation and Development Authority secured funding to preserve and renovate the Western Maryland Railway Station, originally built in 1913. The project also preserved over 100 structures in the adjacent Downtown Cumberland Historic District. Many additional projects are scheduled for completion over the next several years, including development of the Crescent Lawn Festival Grounds, creation of a hiking/biking path connecting the Allegany Highlands Trail to the C&O Canal towpath, private redevelopment of the Footer's Dye Works Building, and rewatering of the western terminus of the C&O Canal. Canal Place is combining historical preservation, recreation, and education with economic development strategies that benefit all of western Maryland.

CONTACT:
Richard Pfefferkorn
Executive Director
Canal Place Preservation and Development Authority
13 Canal Street
Cumberland, Maryland 21502
(301) 724-3655
(800) 989-9394
Email: pfefferkorn@canalplace.org

GOAL TWO PROJECTS

MISSISSIPPI

SHIFTING STRATEGIES TO CREATE MORE JOBS: YELLOW CREEK INLAND PORT INDUSTRIAL SITE

Once planned to support a nearby nuclear power plant, a large rural site in Tishomingo County is now thriving despite several economic setbacks. Although the nuclear plant project was discontinued before completion, and a NASA rocket motor facility on the same site was terminated as well, the port has quietly grown into a valuable economic asset in the area. Recognizing the need to have more control of their economic fate, local leaders created a plan to seek broader-based economic development opportunities by expanding the Yellow Creek State Inland Port Authority's industrial complex and establishing the Northeast Mississippi Waterway Industrial Park. Located at the mouth of the Tennessee River and the Tennessee-Tombigbee Waterway, the port's ability to provide businesses with low-cost transportation for bulky products has attracted the attention of several new industries and potential customers. To date, private industry has invested more than $4 million in the port and created more than 100 high-paying jobs. The Appalachian Regional Commission has provided funding for the water and sewer services and the road improvements necessary to make the private investments possible.

CONTACT:
Eugene Bishop
Executive Director
Yellow Creek State Inland Port Authority
43 County Road 370
Iuka, Mississippi 38852
(662) 423-6088
Email: ycport@network-one.com

EXPANDING TRANSPORTATION OPTIONS TO COMPETE IN THE GLOBAL MARKETPLACE: THE PORT ITAWAMBA MASTER PLAN

As businesses in a 12-county region in northeastern Mississippi and northwestern Alabama search out new global trade opportunities, improving the cost and dependability of accessing international markets is proving critical to their success. First created in 1986 to serve commercial shipping along the Tennessee-Tombigbee Waterway, Port Itawamba is transforming itself from a simple marine terminal to a fully coordinated, multimodal logistic center. As part of a new master plan, the center will offer support services, flexible warehouse space, and efficient container movements via truck, rail, and water. In a recent study of regional transportation issues, the Appalachian Transportation Institute at Marshall University concluded that the availability of a well-balanced, multimodal mix of transportation services, as called for in the Port Itawamba master plan, can enhance access to markets and streamline shipping costs for area businesses by 20 to 25 percent. The new master plan, developed with support from the Appalachian Regional Commission (ARC), presents a carefully conceived growth strategy for Port Itawamba and maps out an important new intermodal transportation network to better serve the needs of area businesses. The Port Itawamba design is one of ten ARC intermodal-planning grant projects launched in the past two years.

CONTACT:
Tim Weston
Director
Port Itawamba
P.O. Box 577
Fulton, Mississippi 38843
(662) 862-4571
Email: icdc@nexband.com

NEW YORK

MAKING SERVICES MORE EFFICIENT: SOUTHERN TIER WEST CENTER FOR LOCAL GOVERNMENT AND COMMUNITY SERVICES

Local governments are providing better services to their communities in Allegany, Cattaraugus, and Chautauqua Counties, thanks to the Southern Tier West Center for Local Government and Community Services. Begun in 1988 as the Community Assistance Program, the center provides local elected officials and employees with a variety of carefully developed training and technical assistance programs designed to improve the way in which they govern and grow their communities. Local governments are also working together to provide more efficient services. Virtually all of the region's 130 local governments are providing more effective and responsive services, thanks to the center's strong advocacy of inter-municipal cooperation. Numerous seminars, workshops, conferences, and other capacity-building programs continue to provide local leaders with the skills and knowledge needed to improve community services. Local financial support for the center's effort continues to grow, with mentorship including 118 out of 130 local governments as of December 2000.

CONTACT:
Eric Bridges
Director
Center for Local Government and Community Services
Southern Tier West Regional Planning and Development Board
Center for Regional Excellence
4039 Route 219, Suite 200
Salamanca, New York 14779
(716) 945-5301
Email: ebridges@southerntierwest.org

CREATING NEW MARKETS FOR FAMILY FARMERS: SOUTHERN TIER SMALL FARM EXPANSION INITIATIVE

Several years ago, as the farming industry continued to consolidate into fewer, larger producers, farm development officials in New York saw great potential for smaller producers in emerging local and urban specialty markets. Unfortunately, many small operators were unaware of the opportunity. In response to this problem, the Southern Tier Small Farm Expansion Initiative provided information and technical assistance to farmers in eight southern New York counties, helping reestablish a demand for grass-fed veal and establish a pastured-poultry industry in the region. The initiative also helped new and existing beef cattle, goat, and sheep producers identify targeted markets. With the assistance of the project, which provides consistent, up-to-date market information to producers, farmers now sell meat products, including meadow-raised veal, in local and regional markets and restaurants. Over 50 farmers now produce pastured poultry, and many have expanded their capacity to meet the growing demand in the new agricultural industry.

CONTACT:
Phil Metzger
Director
Southern Tier Small Farm Expansion Initiative
99 North Broad Street
Norwich, New York 13815
(607) 334-3231 ext. 4
Email: phil.metzger@ny.usda.gov

GOAL TWO PROJECTS

PROVIDING COMMERCIAL ACCESS: PRESCOTT AVENUE INDUSTRIAL ACCESS ROAD

Local officials in Elmira Heights faced a major hurdle in their efforts to attract new businesses to an industrial area that had six major businesses and an older vacated facility ready for redevelopment. The problem was Prescott Avenue, the main road serving the area. The avenue had poor pavement and drainage, no curbing to control storm water, and no pedestrian sidewalks. Without significant road improvements, expansion was impossible. Cooperative funding from the Appalachian Regional Commission, the state, and the village enabled local officials to upgrade Prescott Avenue to meet industrial access road standards. As a result of the improvements, existing businesses expanded and the vacant manufacturing site was developed, securing over 500 jobs for the community.

CONTACT:
Richard Wysowski
Director
Urban Renewal Agency
215 Elmwood Avenue
Elmira Heights, New York 14903
(607) 734-7156
Email: etownhal@stny.rr.com

WORKING TOGETHER FOR PUBLIC HEALTH: RANDOLPH AND EAST RANDOLPH WASTEWATER FACILITIES

Recognizing that poor sewage disposal was threatening the health and economic well-being of many area residents, the small adjacent villages of Randolph and East Randolph in Cattaraugus County decided to work together to resolve the problem. Surveys showed that residences, businesses, and schools in both communities relied on individual septic tanks for sewage disposal. Tight soil and a high water table resulted in the frequent failure of these septic systems. Together the communities hired an engineering firm to design a wastewater system to meet their needs and were able to obtain state and federal funds for construction. The partnership resolved a serious public health problem, eliminated runoff into adjacent waterways, and generated new commercial and residential development.

CONTACTS:
Howard Zollinger
Mayor
Village of Randolph
1 Bank Street
Randolph, New York 14772
(716) 358-9701

Howard Van Rensselaer
Mayor
Village of East Randolph
1 Bank Street
Randolph, New York 14772
(716) 358-6070

NORTH CAROLINA

BUILDING AFFORDABLE HOMES: WESTERN NORTH CAROLINA HOUSING PARTNERSHIP

Leaders from area governments, local development districts, and several area businesses came together in 1988 to develop affordable housing for older adults in Appalachian North Carolina. A nonprofit consortium was formed to help meet the needs of seniors and other special populations. Since then, the partnership has provided technical assistance, housing counseling, and application preparation for those in need. In addition, the group has participated in the development of new housing units. It serves as a general partner on 60 rental units, a service consultant on a 24-unit tax credit/rental production project, a member of a limited liability company on a 48-unit complex for the elderly, and an owner/developer of two eight-unit shared-living residences. It is currently a partner in two additional multifamily developments that will create another 88 units of housing.

CONTACT:
Frank Keel
Director
Western North Carolina Housing Partnership
Isothermal Planning and Development Commission
P.O. Box 841
Rutherfordton, North Carolina 28139
(828) 287-2281
Email: fkeel@regionc.org

OHIO

SAVING JOBS AND PRESERVING A RAIL LINE: AUSTIN POWDER RAIL PROJECT

When CSX announced plans in 1991 to abandon nine miles of rail serving the Austin Powder Company, local leaders in Vinton County were concerned. The powder plant was the county's largest private business, providing more than 260 local jobs. Working closely with Austin Powder officials, community leaders decided to try to save the rail line and sought support from the city of Jackson, in adjacent Jackson County, which already had acquired over 50 miles of track from CSX in an effort to sustain local industry. The city of Jackson secured funding to acquire the Austin Powder line and arranged for the Indiana and Ohio (I&O) short line railroad to operate and maintain the track. In January 1994, the city arranged to transfer track operation and maintenance from I&O to the Great Miami and Scioto Railway Company. Instead of closing, the Austin Powder Company invested $4 million to expand its plant, creating 50 new jobs. The rail acquisition helped stabilize the local economy and maintain rail service to over ten local companies, currently employing over 1,500 people.

CONTACT:
John T. Evans
Mayor
City of Jackson
145 Broadway Street
Jackson, Ohio 45640
(740) 286-2201
Email: jackson@zoomnet.net

GOAL TWO PROJECTS

CREATING NEW INDUSTRIES ON OLD INDUSTRIAL SITES: NEW BOSTON INDUSTRIAL PARK

Many old industrial sites are environmentally contaminated and have come under the federal Superfund program administered by the Environmental Protection Agency (EPA). These properties, called brownfields, are generally not available for any new use until they undergo environmental cleanup, which can be expensive and incur years of delay as technical and legal disputes are litigated. Changes in the Superfund program, however, encourage a level of cleanup sufficient to allow continued or new industrial uses on contaminated sites. The EPA's brownfields initiative promotes cleanup efforts as tools for industrial and economic development, and the New Boston Industrial Park in Scioto County is a model for the reuse of contaminated industrial property. The former location of a steel company, long abandoned, was heavily contaminated. After a portion of the property was decontaminated, a new rail spur encouraged two industrial operations to operate in the park.

CONTACT:
Bob Walton
Executive Director
Community Action
P.O. Box 1525
Portsmouth, Ohio 45662
(740) 354-7541
Email: bwalton@zoomnet.net

CREATING JOBS THROUGH INNOVATION: MEASLEY RIDGE ROAD ELEVATED WATER STORAGE TANK

Although economically distressed, Ohio's Adams County has taken steps to ensure that its largest manufacturer remains successful in producing and marketing red cedar products. With growing sales in the Unites States and abroad, the company wanted to expand its operations at Peebles, in Meigs Township. Other companies had also expressed interest in the area, which is physically attractive and home to residents with a strong work ethic. Unfortunately, the public water system could not meet the demands of existing residences, let alone new industry. The county developed a plan to improve wellfields, water treatment, and water supply. With support from the Appalachian Regional Commission, a new 200,000-gallon elevated composite storage tank will be the first of its kind built in the nation. The tank, along with piping, valves, and fire hydrants, will guarantee a safe water supply and fire protection for previously unprotected homes. This basic infrastructure will help grow the industrial base and improve the economic well-being of the community.

CONTACT:
Brian Ast
Manager
Adams County Regional Water District
P.O. Box 427
West Union, Ohio 45693
(937) 544-2396
Email: acrwd@bright.net

INCREASING WATER CAPACITY FOR A GROWING INDUSTRY: LETART WATER LINE PROJECT

Each year nurseries and greenhouses in the Letart area of Meigs County produce over $5 million in tomato plants, hanging baskets, and flower flats. These products, sold primarily to large national retailers such as Wal-Mart and Kmart, have become a substantial part of the local economy. Recently, however, community officials became concerned about the large quantities of water required by these businesses and the potential water shortage these requirements might engender for farms and residents alike. With support from the Appalachian Regional Commission, Meigs County officials solved the problem by completing the Letart Water Line. With its larger water pipes and a new pumping station, the project has helped ensure that the nursery industry can continue to grow and create new jobs.

CONTACT:
Boyer Simcox
Executive Director
Buckeye Hills–Hocking Valley Regional Development District
Route 1, Box 299D
Marietta, Ohio 45750-0755
(740) 374-9436
Email: bhhvrddmarietta@ee.net

PENNSYLVANIA

BUILDING A WATER LINE TO MAINTAIN JOBS: CUMBERLAND MINE WATER PROJECT

The 500 workers at the Cumberland Mine in Greene County faced the likelihood of layoffs unless mine operations could be expanded to include access to potable water. Mine officials said water shortages threatened operations at two existing mine portals as well as at a proposed third portal. With support from the Appalachian Regional Commission, community leaders approved and built a new ten-mile water line that provided additional water to the mining facility as well as to 51 private residences and several new businesses. The company operating Cumberland Mine agreed to pay monthly user fees, offsetting most debt charges for the project, which included a pump station, a one-million-gallon storage tank, and 45 fire hydrants. Residential user fees also help to offset costs.

CONTACT:
Joseph J. Simatic
General Manager
Southwestern Pennsylvania Water Authority
P.O. Box 2119
Jefferson, Pennsylvania 15344
(724) 883-2301

GOAL TWO PROJECTS

WORKING TOGETHER TO CREATE JOBS: UNION COUNTY BUSINESS PARK

The Union County business community and local government, with the support of state and federal agencies, have successfully launched a new business park with associated residential and recreational development. The 660-acre site was purchased in 1995 by the Union County Industrial Development Corporation, the industrial arm of the Union County Chamber of Commerce. The chamber, Union County commissioners, the state, and several federal agencies have invested more than $18 million toward the project's acquisition and development. The first phase of the park is complete, with over 100 acres available for sale. All lots have public water and sewer service, natural gas availability, and telecommunications capabilities. This initial phase alone is expected to create between 1,000 and 1,500 jobs. The Union County Chamber is now exploring a cooperative effort with the Williamsport/Lycoming Chamber of Commerce to develop an additional 400 acres in Lycoming County adjacent to the current development. If feasibility studies are supportive, a new multicounty development corporation may be established to manage the expanded project.

CONTACT:
Jerry Bohinski
Director of Economic Development
SEDA–Council of Governments
RR #1, Box 372
Lewisburg, Pennsylvania 18737
(570) 524-4491
Email: bohinski@seda-cog.org

IMPLEMENTING A STRATEGY FOR ECONOMIC REVITALIZATION: MEADOW RIDGE BUSINESS PARK

To retain existing businesses and attract new employers, the Greene County Industrial Development Authority, in cooperation with the Regional Industrial Development Corporation of Southwestern Pennsylvania, designed the 108-acre Meadow Ridge Business Park near Interstate 79. The park is an essential component of the Greene County Strategic Plan for economic revitalization. Sixty acres of developed land have been opened under the first phase, which includes construction of a two-lane access road, as well as water and wastewater utilities. The park's first two corporate clients have hired over 85 employees, and the park is planning to open a new facility for additional businesses in the spring of 2002.

CONTACT:
Donald F. Chappel
Executive Director
Greene County Industrial Development Authority
19 South Washington Street, Suite 150
Waynesburg, Pennsylvania 15370
(724) 627-9259
Email: dchappel@greencountyidea.org

TENNESSEE

REPLACING A BRIDGE TO SAVE JOBS: HICKMAN CREEK BRIDGE REPLACEMENT

In 1993, state engineers inspected a deteriorating railroad bridge over Hickman Creek in Putnam County and concluded that the bridge needed to be replaced as soon as possible. Engineers expressed concern about the condition of a 147-foot deck plate girder originally built in 1888. Not only was the bridge's wooden material deteriorating rapidly, but the structure itself was severely disrupting the water flow of Hickman Creek. Putnam County economic development officials also expressed concern. Rail traffic in Putnam County had increased dramatically. Without the bridge replacement, rail service would come to an abrupt halt, putting at risk hundreds of jobs, including many in the area's wood products industry. One local firm, Consolidated Forest Products, alone employed over 100 workers. With support from the Appalachian Regional Commission, the Nashville and Eastern Railroad Authority raised the funds to replace the bridge, saving many local jobs and allowing several local companies to expand.

CONTACT:
Tony Linn
Director
Nashville and Eastern Railroad Corporation
P.O. Box 795
Manchester Center, Vermont 05255
(802) 362-1516
Email: amlinn@nerr.com

IMPROVING PROSPECTS FOR INDUSTRIAL INVESTMENT: GRUNDY COUNTY PELHAM INDUSTRIAL PARK

Recognizing that they must diversify their job base in the face of increased international textiles competition, community officials in Grundy County, Tennessee, embarked on a major initiative to develop a new industrial park adjacent to Interstate 24 in Pelham Valley. With help from state and federal agencies, Grundy County officials acquired land, installed basic infrastructure, and successfully built and sold the initial industrial park building by 1998. In 2001, the county's efforts were again rewarded when Toyo Seat USA broke ground on a new $12 million plant that is expected to employ approximately 200 people. Toyo Seat, which will supply automotive seat frames for a Nissan automobile assembly plant, promises to anchor the business park while benefiting from the strong workforce and pro-business environment in Grundy County. Securing major employers like Toyo Seat has been a major goal for the Pelham Industrial Park. The Appalachian Regional Commission provided support to build a water tank to serve the needs of park occupants. By improving prospects for additional industrial investment and corporate partnerships in the area, the Pelham Industrial Park project in Grundy County could serve as a model for economic revitalization throughout the Appalachian Region.

CONTACT:
Beth Jones
Economic Development Director
Southeast Tennessee Development District
P.O. Box 4757
Chattanooga, Tennessee 37405-0757
(423) 266-5781
Email: bjones@sedev.org

GOAL TWO PROJECTS

VIRGINIA

OVERCOMING MINE-RELATED WATER SAFETY ISSUES: GRUNDY/SLATE CREEK REGIONAL WATER PROJECT

Depending largely on cisterns, wells, and springs for their water, residents of 160 homes outside Grundy in Buchanan County became concerned when mining activities in the area depleted and polluted these resources. County officials sought funding from a variety of sources to design and build a comprehensive water distribution system to correct the problem. In addition to homes near Grundy, the system also serves a number of residences across the border in West Virginia. The recently completed project is expected to improve the quality of life for hundreds of area residents, ensuring a dependable supply of water that meets current safety standards.

CONTACT:
Louis Ballenberger
Senior Planner
Cumberland Plateau Planning District Commission
950 Clydesway Road
P.O. Box 548
Lebanon, Virginia 24266
(540) 889-1778
Email: lbcppdc@naxs.net

EMPOWERING RURAL RESIDENTS TO HELP THEMSELVES: SMITH RIDGE SELF-HELP WATER PROJECT

For 100 years or more, the residents of Smith Ridge—a rural town of about 150 people in Tazewell County—got their water from cisterns, springs, or wells. When the wells ran dry in the summer, some families were forced to haul in fresh water from out of town in a fire truck. Given its small population and remote, mountainous location, the town saw little prospect of improving its water supply. In the summer of 1998, however, the situation changed dramatically as a result of an innovative program that helps people in small towns help themselves. With support from the state's Department of Housing and Community Development, residents banded together and built a seven-mile water-line extension to serve their homes. More than 60 residents, including nearly every able-bodied adult in the town, volunteered to help. As a result, the project cost about $250,000—75 percent less than the estimated contractor cost of over $1 million—and the extension took only three months to complete as opposed to the expected 18 months. Based on the success of this project, the state has supported more than a dozen similar projects in southwestern Virginia, providing water to more than 500 households and saving more than $5 million in construction costs.

CONTACT:
Todd Christensen
Associate Director
Project Management Office
Department of Housing and Community Development
Jackson Center
501 North 2nd Street
Richmond, Virginia 23219-1321
(804) 371-7029
Email: tchristensen@dhcd.state.va.us

WEST VIRGINIA

IMPROVING THE DOWNTOWN AREA: MORGANTOWN WHARF STREET DISTRICT REVITALIZATION

A common vision supported by public and private investment rejuvenated the historic Morgantown Wharf District and made it a vital part of the city named Best Small City in America in 2000. Renewal of this late-nineteenth-century warehouse area adjacent to the downtown business district began in 1997. A new access project connected the district to the Caperton Rail Trail. An old cobblestone main street through the area was rebricked and landscaped. Private developers began looking at the area, renovating old structures and building new ones. By early 2001, a 1,000-seat amphitheater, two restaurants, a train depot, and a bus station had been completed. Eight new businesses, located in six rehabilitated and two new buildings, employed close to 250 workers. Three new buildings, including a hotel, were planned for the coming year. Total private investment committed to the Wharf District was $131 million. These tangible returns on the initial $500,000 public investment—including $250,000 from the Appalachian Regional Commission—will continue for many years. The Wharf District development, originally supported by the City's Vision 2000 Committee, the chamber of commerce, and the Morgantown Area Economic Partnership, continues to be a highly successful public-private collaboration and a model for other communities.

CONTACT:
Dan Boroff
City Manager
City of Morgantown
389 Spruce Street
Morgantown, West Virginia 26505
(304) 284-7434
Email: citymanager@morgantown.com

GOAL THREE

Civic Capacity and Leadership

The people and organizations of Appalachia will have the vision and capacity to mobilize and work together for sustained economic progress and improvement of their communities.

ALABAMA

BUILDING COMMUNITY STRENGTH BY PRESERVING THE PAST: ALICEVILLE MUSEUM AND CULTURAL ARTS CENTER

When the Appalachian Regional Commission–sponsored Aliceville Downtown Revitalization Project brought diverse representatives of the community together, they embarked on an ambitious project to preserve unique aspects of Pickens County history. The Aliceville Museum and Cultural Arts Center was the local community's creation, realized step-by-step with hands-on help from area residents. A mural based on a 1944 sketch by a German prisoner of war commemorates the community's unique role in World War II. Other exhibits honor Pickens County veterans from the Revolutionary War to the present. Farm equipment and antiques recall a life on the farm fast fading from view. The center is already seeing positive results from these efforts: museum attendance and membership continue to grow.

CONTACT:
Mary Bess Paluzzi
Director
Aliceville Museum and Cultural Arts Center
104 Broad Street Plaza
Aliceville, Alabama 35442
(205) 373-2363
(888) 751-2340
Email: museum@pickens.net

DEVELOPING JOB TRAINING TO MEET INDIVIDUAL NEEDS: SCOTTSBORO APPALACHIAN COMMUNITY LEARNING PROJECT

In 1993, a chamber of commerce survey found that 42 percent of Jackson County's adults had not earned a high school diploma, and 17 percent of the adults were functionally illiterate. In response, business and community leaders formed the 21st Century Council to promote adult education with a special emphasis on job training. Staff and volunteers at the council's Adult Career Center have had success working one-on-one with individuals seeking employment training. By focusing on the barriers to employment unique to each client and working extensively with employers and social service agencies, the center has helped over 100 people find work. The program has won the praise of clients—many of whom were formerly on welfare—and has had strong support from local business and community groups.

CONTACT:
Ann Kennamer
Executive Director
21st Century Council
305 South Scott Street, Suite 21
Scottsboro, Alabama 35768
(256) 218-2121
Email: career21@mail1.scottsboro.org

GOAL THREE PROJECTS

WORKING TO BRING HOME THE BEST AND THE BRIGHTEST: HALE BUILDERS OF POSITIVE PARTNERSHIPS PROGRAM

Concerned that many high-achieving high school students do not return to live in Hale County, Alabama, after college, the county's Hale Empowerment and Revitalization Organization Family Resource Center created a youth leadership program to encourage young people to build personal and professional lives in this distressed area of western Alabama. Over 20 student leaders from county high schools were chosen as the first participants in the Hale Builders of Positive Partnerships (BOPP) program. With support from the Appalachian Regional Commission, these "Hale BOPP Comets" received training in leadership and business skills, learned about local history and culture, and took part in local service projects. The University of Alabama, Auburn University, and Shelton State Community College cooperated in activities that will build problem-solving skills, encourage responsibility, and reveal the value of long-term community commitment. The program has been so successful that it has now expanded to include additional training and educational opportunities for the Comets during their senior year of high school and continued contact after they graduate and proceed to college.

CONTACT:
Jim Kellen
Director
HERO Family Resource Center
1015 Market Street
Greensboro, Alabama 36744
(334) 624-9100
(888) 444-4376
Email: jwkellen@yahoo.com

FOSTERING COMMUNITY DEVELOPMENT THROUGH LEADERSHIP TRAINING: YOUR TOWN ALABAMA PROGRAM

This leadership development and training program is providing strong new networks across government, business, and local groups in Alabama communities. The Your Town program conducts two workshops a year targeted at the community planning and design problems that affect rural Alabama. Increasing urbanization provides opportunities and challenges, and the program teaches participants about strategic planning and decision making at the community level. A statewide board of directors is working with partners including the Auburn University Center for Architecture and Urban Studies, the University of Alabama Center for Economic Development, and the Regional Planning Commission of Greater Birmingham to refine the curriculum, develop a continuing education program for alumni, and establish a youth program for high school students. Partnership has been part of Your Town from its inception. The National Trust for Historic Preservation initially launched the program, and the Appalachian Regional Commission has funded curriculum development and workshops. The board of directors represents a wide range of public- and private-sector organizations and interests. With this strong public and private support for Your Town, the board expects the program to become a self-sustaining, ongoing resource for Alabama.

CONTACT:
Paul Kennedy
Project Coordinator
Cawaco RC&D Council
2112 Eleventh Avenue South, Suite 220
Birmingham, Alabama 35205
(205) 251-8139 ext. 35
Email: paul.kennedy@al.usda.gov

GEORGIA

USING TECHNOLOGY TO ENHANCE COMMUNITY DEVELOPMENT: GROWTH MANAGEMENT INITIATIVE WEB SITE

A 14-county area of northwestern Georgia in which the population is expected to double in the next 15 years has initiated an effort to develop a common regional vision and strategy for managing growth and change. To be successful, such a strategy must be embraced by businesses, local government officials, and residents throughout the region and be reflected in their decisions. The counties want the initiative to be a positive force for enhancing community development, not the focus of a "zoning or no zoning" or "anti-growth" debate. The Internet is proving an excellent way to inform participants, raise issues, and provide a forum for broad public comment on growth and development concerns. The Growth Management Initiative Web site contains an online survey and a feedback section where comments on proposals or new issues can be made and where tools for proactive growth management can be made available to local leaders.

CONTACT:
Leamon Scott
Regional Representative
Georgia Department of Community Affairs
527 Broad Street
Rome, Georgia 30161
(706) 802-5490
Email: lscott@dca.state.ga.us

KENTUCKY

SYNCHRONIZING LOCAL PLANNING AND DEVELOPMENT RESOURCES: KENTUCKY APPALACHIAN COMMUNITY DEVELOPMENT INITIATIVE

The Kentucky Appalachian Community Development Initiative (CDI) encourages local strategic planning through priority access to state development resources. Municipalities are encouraged to cooperate with county government and citizens in creating a community development concept plan that defines goals to be achieved within ten years. Communities competitively selected for participation receive enhanced planning services and other technical assistance to create a strategic action blueprint to meet those goals. In addition, they receive priority consideration for grant awards and state assistance to fulfill that action blueprint. For example, two communities, Hindman (in Knott County) and Jenkins (in Letcher County), were the first selected in the program after it was launched in 1997. Both communities are now pursuing long-term development strategies to restructure their economies, create employment opportunities, and enhance the quality of life they offer. Two additional communities, Paintsville and a rural area that includes

GOAL THREE PROJECTS

Benham, Cumberland, and Lynch, have launched similar programs. While the CDI emphasizes community-based planning, it also focuses on civic leadership, physical infrastructure and workforce, and the community's history of financial responsibility and current fiscal capacity to support development initiatives. By better synchronizing local strategic planning with development resources, CDI speeds up economic development while using state resources more efficiently.

CONTACT:
Ewell Balltrip
Executive Director
Kentucky Appalachian Commission
Gorman Center
601 Main Street, Suite 001
Hazard, Kentucky 41701
(606) 435-6129
Email: ewell@mis.net

GIVING PEOPLE POSITIVE OPTIONS: HAZARD PERRY COUNTY COMMUNITY MINISTRIES, INC.

Hazard Perry County Community Ministries, Inc. (HPCCM), is a grassroots, community-based nonprofit organization that has generated more than $10 million in ongoing and sustainable services over the past ten years. HPCCM was first incorporated in 1976 and is governed by a volunteer board of local leaders and constituents who take an active role in program development, fundraising, and community awareness. A guiding premise is that given good options, people will make good choices. The organization's mission is to analyze resources, identify gaps, and determine how best to lead the community in human, economic, and community development. Many programs, including large capital projects, have been developed, and all have been successful and self sustaining. Among these are child and adult day care, homelessness intervention and emergency shelter provision, home-ownership promotion, economic development, adult education, budget and housing counseling, and downtown revitalization. A variety of public- and private-sector partners help make these programs an ongoing success.

CONTACT:
Gerry F. Roll
Executive Director
Hazard Perry County Community Ministries, Inc.
P.O. Box 1506
Hazard, Kentucky 41702
(606) 436-2662
Email: hpccmadm@mis.net

MARYLAND

MANAGING FOR SUCCESS: REGIONAL MAIN STREET PROGRAM

Fearing that their downtown business redevelopment efforts might be piecemeal and haphazard without consistent managerial leadership, two western Maryland towns have engaged full-time Main Street managers. The managers in Cumberland (Allegany County) and Oakland (Garrett County) initiate and coordinate a variety of administrative, management, and promotional activities. Working closely with volunteer organizations, as well as business tenants and property owners, they are full-time advocates and sources of information on downtown facilities, programs, and opportunities. The managers help both downtown revitalization programs avoid the burnout and inconsistency that plague many volunteer organizations, while bringing new visibility, activity, and commerce to Cumberland and Oakland. The program has proven so successful that Frostburg has joined the program and plans to hire its own Main Street manager in the fall of 2001.

CONTACTS:
Ed Mullaney/Sue Cerutti
Downtown Managers
City of Cumberland
P.O. Box 1702
Cumberland, Maryland 21502
(301) 777-2800
Email: cu_dw_dev@allconeet.org

Glenn Tolbert
15 South Third Street
Oakland, Maryland 21550
(301) 533-4470

NEW YORK

BUILDING CAPACITY FOR ECONOMIC DEVELOPMENT: STC PARTNERSHIP PROGRAM

Partnerships among community groups, governments, and local professionals of different disciplines are required to successfully design and sustain community economic development projects. As relationships are formed, communities develop greater capacity to enhance growth and meet community needs. Working with a task force of local professionals and support from the Appalachian Regional Commission, the program provides approximately 225 hours of in-depth technical assistance to communities with identified local needs and projects. Municipalities contribute $5,000,

GOAL THREE PROJECTS

designate a team of local leaders, and agree to develop and implement a detailed scope of work. Projects have included water and sewer studies, downtown revitalization initiatives, and the development of requests for proposals to meet defined community needs. Because the project requires intensive time commitments, only four communities are selected to participate each year. The project advisory committee plans to ensure sustainability by integrating the program into existing community and local government structures, encouraging multi-community partnerships, expanding technical assistance services for nonprofit agencies, and developing ways that will insure the program continues to be available to smaller communities with short-term project needs.

CONTACT:
Tom McGarry
Community Development Specialist
Southern Tier Central Regional Planning and Development Board
145 Village Square
Painted Post, New York 14870
(607) 962-5092
Email: tmcgarry@stny.rr.com

NORTH CAROLINA

BUILDING LOCAL LEADERSHIP: NORTH CAROLINA RURAL ECONOMIC DEVELOPMENT INSTITUTE
Although rural North Carolina enjoys a rich tradition of community involvement and civic participation, leaders of many rural communities are being asked to face unprecedented economic challenges. To prepare a broad, diverse group of rural leaders with the knowledge and skills necessary to manage economic transitions in their communities and promote sustained development, the North Carolina Rural Economic Development Center established its first Rural Economic Development Institute in 1989. Today, over 560 people have graduated from the program. Participants in the institute come from a broad range of professionals and volunteers who are selected based on their applications for admission. Participants complete three rigorous instruction sessions that focus on improving people skills, learning the building blocks of successful economic development, and implementing strategies—the "how to's"—including strategic planning, coalition building, and conflict management. The program's alumni form a mentoring network for the new graduates.

CONTACT:
Robin Pulver
Vice President
Community and Human Resources Development
North Carolina Rural Economic Development Center
4021 Carya Drive
Raleigh, North Carolina 27610
(919) 250-4314
Email: pulver@ncruralcenter.org

DEVELOPING PRACTICAL APPROACHES TO ECONOMIC DEVELOPMENT: SMALL COMMUNITIES RURAL LEADERSHIP INITIATIVE

Many small communities in western North Carolina lack the professional staff and large leadership base needed to develop a sustainable economy. The Small Communities Rural Leadership Initiative, established and coordinated by HandMade in America, develops practical new approaches to economic development. Participants learn a systematic approach to managing projects, building a leadership base, and involving the broader community in an inclusive, collaborative process. While taking part in training sessions over a full year and using their new leadership skills to recruit others, participants plan and conduct local community projects with assistance from the initiative. Training and project activities help create a long-term leadership corps and encourage cooperation between participating towns. In addition, the models developed in these towns are being made available to communities throughout 21 western North Carolina counties. In a four-year period, these towns have initiated over $17 million in main street revitalization projects.

CONTACT:
Rebecca Anderson
Executive Director
HandMade in America
P.O. Box 2089
Ashville, North Carolina 28802
(828) 252-0121
Email: wnccrafts@aol.com

OHIO

PROVIDING HANDS-ON TRAINING AT THE LOCAL LEVEL: CORPORATION FOR OHIO APPALACHIAN DEVELOPMENT

Based in Athens County, the Corporation for Ohio Appalachian Development (COAD) is a private, nonprofit organization representing 17 community action agencies serving 30 counties. COAD provides a forum for collaboration and coordination among local providers who assist the elderly and low-income individuals and families. COAD also seeks to improve the leadership and management capabilities of local governments and organizations serving local

GOAL THREE PROJECTS

communities. COAD's Appalachian Leadership Academy is a hands-on training program for middle-management staff, giving professional development opportunities to community action, economic development, local government, and nonprofit agency professionals.

CONTACTS:
Roger McCauley
Executive Director
Corporation for Ohio Appalachian Development
P.O. Box 787
Athens, Ohio 45701-0787
(740) 594-8499
Email: rmccauley@coadinc.org

Karen Collins
Director
Appalachian Leadership Academy
Corporation for Ohio Appalachian Development
P.O. Box 787
Athens, Ohio 45701-0787
(740) 594-8499
Email: kcollins@coadinc.org

INCREASING PHILANTHROPIC SUPPORT FOR LOCAL COMMUNITIES: THE FOUNDATION FOR APPALACHIAN OHIO

The Foundation for Appalachian Ohio was established in 1998 as a regional community foundation to address the needs and opportunities of the 29 counties of Appalachian Ohio. The foundation seeks to build a better future for the region by building charitable endowments for regional and local grantmaking. The core business is designed to help donors achieve their charitable goals and connect the people who care with the issues that matter in Appalachian Ohio. As a steward of charitable gifts, the foundation is a regional vehicle through which permanent, charitable endowment is attracted and invested in Appalachian Ohio.

CONTACT:
Leslie Lilly
President and CEO
The Foundation for Appalachian Ohio
P.O. Box 456
Nelsonville, Ohio 45764
(740) 753-1111
Email: llilly@ffao.org

PENNSYLVANIA

PROVIDING A VISION FOR LOCAL LEADERSHIP: GREENE COUNTY STRATEGIC PLAN

The Greene County Strategic Plan recognizes that helping a distressed area help itself requires preparing local civic leadership and institutions to meet social, educational, and economic needs. Created in 1997, the plan includes a community leadership vision for economic development; a strength, weakness, opportunity, and threat analysis of the county; a specific action plan for industry retention, incubation, and expansion; a priority assessment of sites and infrastructure; an analysis of the county's resources to implement the plan; and a detailed workplan and timeline. By recognizing and working to meet the central requirement for civic leadership, Greene County officials are well on their way to achieving the plan's goals and timelines, while increasing private-sector confidence and investment.

CONTACT:
Ann Bargerstock
Director of Planning and Development
Greene County
93 East High Street, Room 220
Waynesburg, Pennsylvania 15370
(724) 852-5300
Email: annie@county.greenepa.net

SOUTH CAROLINA

CREATING NEW INFORMATION TOOLS TO CREATE JOBS IN SOUTH CAROLINA: INFOMENTUM

InfoMentum provides integrated research tools designed to attract industry and create jobs in South Carolina's six Appalachian counties. An unprecedented partnership of county governments, utilities, and the Appalachian Council of Governments (ACOG), assisted by the Appalachian Regional Commission, began planning this standardized information system in 1996. Geographic information systems (GIS) technology provides multiple layers of maps and data, allowing users to generate localized reports, maps, and graphs. A Fact Finder database contains statistics for 15 categories of socioeconomic, population, and quality-of-life attributes. Available industrial buildings, sites, and parks are also included, and all databases are updated monthly. Users can also request special reports, research information, and technical assistance. In this unique collaborative effort, now funded annually by public and private investment, ACOG offers technical support, database design and maintenance, training, administration, and marketing. By providing consistent regional information for all the counties, InfoMentum is a catalyst for fostering regional data partnerships and coordinating community GIS and planning activities. The InfoMentum program is now fully integrated into the region's economic development and planning activities, enhancing the capacity of local government while bringing new development to the region as a whole.

CONTACT:
Carol Andersen
Program Coordinator
Appalachian Council of Governments
P.O. Drawer 6668
Greenville, South Carolina 29606
(864) 242-9733
Email: andersen@scacog.org

DEVELOPING A BLUEPRINT FOR SUCCESS: APPALACHIAN REGIONAL STRATEGIC PLAN FOR ECONOMIC DEVELOPMENT, 2000–2005

Broad community involvement is the key to successful design and subsequent implementation of South Carolina's ambitious Appalachian Regional Strategic Plan for Economic Development. The initiative brings together citizens, economic development leaders, elected officials, service providers, and business representatives to enhance the economic development potential of the region. Participants identify the most critical economic development issues facing the region, develop strategies for addressing issues, and work to implement those strategies. The process builds on a successful regional strategic planning process undertaken by the Appalachian Council of Governments and the South Carolina Department of Commerce between 1992 and 1994. A key principle of the current initiative holds that a knowledgeable and motivated citizenry provides a critical resource for economic development. By involving large groups of citizens, the process builds an understanding of the value of development and the resources necessary to attract industry. It focuses on a vision for development and builds a broad base of support for given strategies. Organizers expect to receive close to 50 specific recommendations that will identify tasks, those who will complete them, the costs, and project deadlines. Since those responsible for implementation are now involved in developing the recommendations, organizers expect this partnership to result in commitment, action, and results.

CONTACT:
Steve Pelissier
Planning Director
Appalachian Council of Governments
P.O. Drawer 6668
Greenville, South Carolina 29606
(864) 242-9733
Email: stevep@scacog.org

TENNESSEE

RECOGNIZING COMMUNITY EXCELLENCE: THE GOVERNOR'S THREE-STAR PROGRAM

For the past 20 years, the Governor's Three-Star Program has helped Tennessee communities preserve and create new employment opportunities, increase family incomes, improve the quality of life, and create a strong leadership base. The program provides recognition and support to communities as they develop, implement, and enhance community development efforts. In order to receive and maintain Three-Star certification, a community must meet basic requirements in organizational, community, educational, economic, and workforce development. Communities are encouraged to undertake additional proposed activities within these goal areas. A team of economic and community development professionals evaluates each community's progress annually. Sixty-two communities qualified for 2000 Three-Star certification; close to half of them are in Appalachian Tennessee.

CONTACT:
Jimmy Earle
Assistant Commissioner for Community Development
Tennessee Department of Economic and Community Development
William R. Snodgrass Tennessee Tower, Tenth Floor
312 Eighth Avenue, North
Nashville, Tennessee 37243-0405
(615) 741-2373
Email: jearle@mail.state.tn.us

GOAL FOUR

Dynamic Local Economies

Appalachian residents will have access to financial and technical resources to help build dynamic and self-sustaining local economies.

ALABAMA

PREPARING AND MARKETING SPECIALTY FOODS: THE SHOALS COMMERCIAL CULINARY CENTER

When four local entrepreneurs expressed interest in sharing a commercial kitchen, the Shoals Entrepreneurship Center conducted a survey and found that more than 40 existing or potential small businesses were seeking access to such a facility. None existed in Alabama, and food preparation companies had to contract their food production to distant locations. The center contacted the Tennessee Valley Authority, which agreed to donate kitchen equipment from its Muscle Shoals Reservation, while the Appalachian Regional Commission and the U.S. Department of Agriculture provided additional funding to develop a culinary incubator. The Florence City School Board made cafeteria and dining space available at its community education facility, where special-foods producers, caterers, and cart vendors now make a wide variety of food products. Currently, the Shoal Commercial Culinary Center, which can accommodate seminars and business conferences, provides training support in pricing, marketing, distribution, and bulk purchasing and plans to offer training in production and packaging, regulatory requirements, product stability, and restaurant presentation when it receives additional funding.

CONTACT:
Jerry Davis
Executive Director
Shoals Entrepreneurial Center
3115 Northington Court
Florence, Alabama 35630
(256) 760-9014
Email: sshaw@shoalsec.com

HELPING LOCAL COMPANIES COMPETE WORLDWIDE: PROJECT TEAM TRAINING PROGRAM

Wood products, apparel, and processed food companies have accounted for over 40 percent of Alabama's manufacturing jobs in recent years, but only about 13 percent of the state's exports. Small and medium-sized Appalachian firms in these industries were initial candidates for the Targeted Export Assistance and Management (TEAM) export assistance project at the University of Alabama's International Trade Center. Many strong, successful companies were interested in exporting but were intimidated by basic questions, such as how to ship, how to receive payment, or how to respond to foreign price queries. Currently, companies with the organizational and financial capabilities to be successful exporters make a commitment to the program and in return receive hands-on assistance in entering a foreign market. In 1997, the program's first year, ten firms exported over $3 million worth of goods. The program continues to add new firms, helping expand markets for business and employees. Now Alabama-made cookies are for sale in Israel, and Alabama food seasonings are sold in Mexico.

CONTACT:
Brian Davis
Associate Director
Alabama International Trade Center
Box 870396
Tuscaloosa, Alabama 35487-0396
(205) 348-7621
Email: Bdavis@aitc.ua.edu

GOAL FOUR PROJECTS

MOVING FROM WELFARE TO WORK: AUBURN UNIVERSITY MONTGOMERY MENTORING ENTREPRENEURIAL NETWORK FOR WOMEN

At least 1,500 women in Appalachian Alabama are 30-month welfare recipients and face daunting challenges in moving from welfare to work. Those located in three economically distressed counties have a more difficult transition as local jobs are limited. To assist women in Hale, Macon, and Pickens Counties, Auburn University Montgomery (AUM) initiated a mentoring program for potential entrepreneurs. Successful women entrepreneurs who have expertise in working with disadvantaged women, owning and operating their own businesses, and working with economic development programs were recruited as core mentors and matched with 15 potential women entrepreneurs. The 15 potential entrepreneurs meet regularly with their core mentors to address issues of importance, including technology use, business plan development, marketing, and business organization. AUM expects the network to expand over time as successful program graduates become mentors to future women entrepreneurs within their respective communities.

CONTACT:
Darla Graves
Project Manager
Alabama Mentoring Entrepreneurial Network
Auburn University Montgomery
Center for Business and Economic Development
600 South Court Street, Suite 110
Montgomery, Alabama 36104
(334) 244-3701
Email: DGraves@CBED.AUM.edu

GEORGIA

PROVIDING SMALL LOANS TO CREATE BIG CHANGES: THE MOUNTAIN PARTNERSHIP LOAN FUND

The Mountain Partnership Loan Fund provides loans of up to $5,000 for individual entrepreneurs or very small businesses in 12 northeastern Georgia counties. Appalachian Community Enterprises, Inc., a nonprofit community-based organization, created the revolving microloan program for low-income or at-risk entrepreneurs who do not qualify for conventional loans but are willing to receive business education and technical support with their loans. Partnerships with area businesses and educators provide the support structures that encourage the financial success of loan recipients.
The Community Bank and Trust of Cornelia processes loans and collections; members of the northeastern Georgia business community review loan applications and serve as mentors and consultants; and the University of Georgia Small Business Development Center provides technical assistance and free business counseling, as does the North Georgia College and State University. Loan-fund orientations and ongoing business resources are provided by North Georgia Technical College through their entrepreneurial education certificate program.

CONTACT:
Grace Fricks
Loan Fund Manager
Mountain Partnership Loan Fund
1727 Turner's Corner Road
Cleveland, Georgia 30528
(706) 348-6609
Email: fricks@alltel.net

CREATING OPPORTUNITIES THROUGH TRAINING: NORTH GEORGIA COLLEGE AND STATE UNIVERSITY'S GEORGIA APPALACHIAN DEVELOPMENT CENTER

Internet marketing and e-commerce are increasing tourism and small business growth in Appalachian northern Georgia. The Georgia Appalachian Development Center provides educational and training support in business and technology, with a focus on Web development, marketing, and e-commerce. The center works closely with the Intellectual Capital Partnership Program at the university, and together the two programs have built a state-of-the-art computer lab for high-tech training. Small businesses learn how to increase their market share and sales through the Internet. Community-based agencies learn how to use the Web to market their city, county, and region. In its first year, the center served 479 people in classes, 41 in financial-resources seminars, and over 140 in technology workshops. The center's partnerships include support from area chambers of commerce, tourism agencies, and development authorities.

CONTACT:
Kimberly R. Foster
Director of Public Services
Georgia Appalachian Development Center
North Georgia College and State University
Dahlonega, Georgia 30597
(706) 867-2814
Email: kfoster@ngcsu.edu

HELPING SMALL BUSINESSES FIND GLOBAL OPPORTUNITIES: GORDON COUNTY INTERNATIONAL EXPORT PROGRAM

The Gordon County Chamber of Commerce is helping small and medium-sized enterprises (SMEs) take advantage of global business opportunities. The Gordon County International Export Program has created an international help desk and clearinghouse for access to federal, state, local, and commercial organizations and Web sites; organized a global trade conference in partnership with local, state, educational, and business organizations; identified local experts as mentors; established partnerships between the chamber and university and state organizations to sustain programming and services; and trained chamber staff in SME export needs and information resources. As a result, four county enterprises are pursuing export opportunities, and the chamber is planning a future program on international trade in collaboration with the University of Georgia's Small Business Development Center.

CONTACT:
Jimmy Phillips
President
Gordon County Chamber of Commerce
300 South Wall Street
Calhoun, Georgia 30701
(706) 625-3200
Email: jphillips@gordonchamber.org

GOAL FOUR PROJECTS

KENTUCKY

HELPING WOMEN ENTREPRENEURS: WOMEN'S INITIATIVE NETWORKING GROUPS

To help more women in Appalachia go into business for themselves, the Appalachian Regional Commission supports a range of programs, including entrepreneurial training, marketing consultations, networking, and mentoring opportunities for low- and moderate-income women. In eastern Kentucky, the Commission has supported the Women's Initiative Networking Groups, a nonprofit organization offering business and marketing assistance to low- and moderate-income women. The program has helped launch over 72 businesses ranging from traditional ventures like child care, tailoring, or craft operations to more innovative enterprises in agriculture, health, and technology.

CONTACT:
Jeannie Brewer
Executive Director
WINGS
433 Chestnut Street
Berea, Kentucky 40403
(859) 985-9753
Email: jbrewer@wingsnet.org

MARYLAND

THINKING BIG WITH SMALL LOANS: WESTERN REGION SMALL BUSINESS DEVELOPMENT CENTER REVOLVING MICROLOAN FUND

A revolving microloan fund has been established to encourage community business investment in Allegany and Washington Counties, providing prime interest rate loans of $10,000 or less for qualifying small business start-ups or expansions. Loan recipients are individuals who cannot qualify for traditional business loans even though their budding businesses show promise. In addition to a loan, these recipients receive technical assistance and counseling to help them manage their new or expanding enterprises. The initiative has already helped close to 20 new businesses to get off the ground. The microloan fund began in 1998 with initial capitalization of $80,000 and was recapitalized in 2000 with an additional $90,000. As principle and interest are paid back, the fund will continue to serve the local small business community.

CONTACT:
Sam LaManna
Executive Director
Western Region Small Business Development Center
957 National Highway, Suite 3
LaVale, Maryland 21502
(301) 729-2400
Email: lamanna@sbdc-wmd.com

HELPING START AND EXPAND LOCAL BUSINESSES: TRI-COUNTY COUNCIL FOR WESTERN MARYLAND REVOLVING LOAN FUND

When this new revolving loan fund was set up in 1985, no one knew for sure how great the need would be or how successful newly funded companies would become. In the past 14 years, 49 business start-ups and expansions have been funded in the manufacturing, retail, construction, and food-service industries. Approximately $2.9 million in loans have leveraged another $18.9 million of private investment in Allegany, Garrett, and Washington Counties. Almost 600 jobs have been created or preserved, generating continuing income and economic growth in the region.

CONTACT:
Leanne Mazer
Executive Director
Tri-County Council for Western Maryland, Inc.
111 South George Street
Cumberland, Maryland 21502
(301) 777-2158
Email: lmazer@tccwmd.org

SUPPORTING NEW ENTREPRENEURS: MICROWORKS INITIATIVE

Business loans are just part of the help available to make small new businesses in Garrett County a success. Under the MicroWorks entrepreneurial initiative, individuals who want to start or expand a small business can take classes and receive technical assistance in cooperation with the Western Maryland Small Business Development Center and the Continuing Education Department at Garrett Community College. In addition, one-on-one consulting and service assistance are available for financial counseling, tax assistance, marketing, promotional assistance, and other services. Links are made where needed with social services assistance, and support is offered to overcome individual barriers. To date, 24 loans averaging $12,000 have been made from the revolving load fund, leveraged with federal, state, and private dollars. Another innovative strategy facilitated by MicroWorks is a partnership with the Mountain Arts Cooperative, which enables Appalachian artisans to work in a cooperative environment and benefit from training, peer exchange, and increased market access. Networking and peer counseling are both available through the Garrett County Chamber of Commerce, another partner in the comprehensive effort to increase entrepreneurial activity and success.

CONTACT:
Glenn Tolbert
Vice President for Community and Economic Development
MicroWorks
Garrett County Community Action Committee, Inc.
104 East Center Street
Oakland, Maryland 21550
(301) 334-9431 ext. 153
Email: mworks@garrettcac.org

GOAL FOUR PROJECTS

PROVIDING GREATER INTERNET ACCESS: WMDNET

Beginning with initial efforts to provide distance-learning opportunities to western Maryland high school students, WMDnet helped develop Internet access for the public sector in three counties. It stimulated entry of private Internet service providers into the region and has been the focal point of various computer and telecommunications projects that benefit students and improve the use of public information. In Hagerstown, a fiber-optic network links numerous agencies and the Internet. In Allegany County, Allconet links agencies, schools, nonprofits, and colleges while providing Internet access. The system has successfully demonstrated Internet linkages via a wireless hookup and is building now to expand those services across the county. In Garrett County, the Garrett Rural Information Cooperative provides Internet access to private and public sectors. The Western Maryland Internet Lab at Frostburg State University is one of a number of associated initiatives that continue to evolve from this ongoing work.

CONTACT:
Frank Peto
Director
Regional Education Service Agency
127 South Smallwood Street
Cumberland, Maryland 21502
(301) 777-3525
Email: fpeto@allconet.org

MISSISSIPPI

IDENTIFYING A NEW POOL OF WORKERS FOR RURAL BUSINESSES: THE CREATE FOUNDATION REGIONAL LABOR ANALYSIS STUDY

Recognizing that most new employers want to hire experienced workers, officials in northeastern Mississippi decided to take a different approach in examining the area's workforce. Going beyond basic employment figures, local leaders sought to identify the number of local workers interested in upgrading to new and better-paying jobs. The result was the CREATE Foundation Regional Labor Analysis Study, which offered important new information on local workers interested in moving up the job ladder. The study found that these local individuals make up the best pool of potential workers for new and expanding business developments in northeastern Mississippi. By describing and documenting the existence of these workers, the study has become an important tool for local officials to use in attracting new businesses to the region.

CONTACT:
Morgan Baldwin
Director of Programs
CREATE Foundation
P.O. Box 1053
Tupelo, Mississippi 38802
(662) 844-8989
Email: morgan@createfoundation.com

NEW YORK

FINANCING FOR NEW SMALL BUSINESSES: REGIONAL ECONOMIC DEVELOPMENT AND ENERGY CORPORATION REVOLVING LOAN FUND

Even in a booming economy, local lending institutions are often reluctant to finance small business start-ups. Government-sponsored revolving loans fill a critical financing gap in many areas, such as Chemung, Schuyler, and Steuben Counties, where a regional revolving loan fund has been helping finance business development projects since 1980. The Appalachian Regional Commission (ARC) provided $1.3 million in initial capital, and over the next 20 years, the fund disbursed over $4.9 million in loans. These loans leveraged over $38 million in private investment, principally for industrial manufacturing and commercial businesses. By working closely with other agencies and private lenders and by requiring non-ARC funds for at least 50 percent of project financing, the program has kept the success rate for funded projects high. Thirty-one of 33 enterprises funded in the past five years are still in operation, employing over 400 people.

CONTACT:
Diane W. Lantz
Executive Director
Regional Economic Development and Energy Corporation
145 Village Square
Painted Post, New York 14870
(607) 962-3021
Email: redec@stny.rr.com

USING NEW TECHNOLOGIES FOR REGIONAL MARKET PLANNING: SOUTHERN TIER WEST REGIONAL INTERNET-BASED GEOGRAPHIC INFORMATION SYSTEM

Developers and companies seeking sites for new facilities must analyze topographic information, existing utility and transportation infrastructure, and various jurisdictional issues including zoning requirements. A new, Internet-based system makes that analysis much easier and also provides a set of tools that can be used to market, study, and advance the three-county Southern Tier West region of New York. The geographic information system (GIS) creates, assembles, and organizes layers of maps, including parcel boundaries, zoning, municipal water and sewer coverage, wetlands, flood plains, and other topographical details. The GIS data include tax rates, natural gas hookups, electric service hookups for existing buildings, and other infrastructure hookup information. Approximately 250 maps are generated each day, and 5 percent of site visitors request additional information. The Southern Tier West Regional Planning and Development Board took the lead in the project, in partnership with local governments and community organizations. The system has already fostered the development and revision of strategic plans throughout the region.

CONTACT:
Brian Schrantz
Director of Information Services
Southern Tier West Regional Planning and Development Board
4039 Route 219, Suite 200
Salamanca, New York 14779
(716) 945-5301
Email: bschrantz@southerntierwest.org

GOAL FOUR PROJECTS

NORTH CAROLINA

PARTNERING FOR SUCCESS: SOUTHWESTERN NORTH CAROLINA REVOLVING LOAN FUND

A comprehensive partnership with supporting institutions ensures success for this revolving loan fund, administered by the Southwestern North Carolina Planning and Economic Development Commission. The commission works closely with its partners in a seven-county region, including business and technology centers at area colleges, the state department of commerce, commercial banks and lenders, and nontraditional lenders, including the Self-Help Credit Union and the Mountain Microenterprise Fund.

CONTACT:
Vicki Greene
Revolving Loan Administrator
Southwestern North Carolina Planning and Economic Development Commission
P.O. Drawer 850
Bryson City, North Carolina 28713
(828) 488-9211
Email: vicki@regiona.org

OHIO

CREATING HOMEGROWN JOBS: OHIO APPALACHIAN HARDWOOD INITIATIVE

Appalachian hardwood products are sparking new economic development efforts, from local skills training in rural Ohio counties to international marketing in Europe and Asia. The Appalachian Regional Commission has helped fund technical assistance for potential entrepreneurs, including workshops, manuals, and demonstrations, through the Ohio State University Research Foundation. Workshop participants not only learn how to fashion products from oak, hickory, walnut, cherry, hard maple, ash, and yellow poplar, but also how to reach current and potential markets for such products. The Ohio Valley Regional Development Commission is also helping existing manufacturers grow markets for Appalachian hardwood products through international trade shows in Germany and Japan. Participating companies join a regionally clustered trade delegation and receive cost sharing (of not more than 50 percent) to pay for tradeshow exhibit space. The network of support for hardwood products development includes the state, the university, individual companies, and other federal agencies, including the Market Development Cooperators Program of the U.S. Department of Commerce.

CONTACTS:
John Hemmings
Ohio Valley Regional Development Commission
P.O. Box 728
Waverly, Ohio 45690-0728
(740) 947-2853
(800) 223-7491
Email: jhemmings@ovrdc.org

Douglas Fry
Business Development Specialist
Ohio Valley Regional Development Commission
P.O. Box 728
Waverly, Ohio 45690
(740) 947-2853
(800) 223-7491
Email: dfry@ovrdc.org

HELPING A WOMAN-OWNED BUSINESS EXPAND: REVOLVING LOAN FUND ASSISTANCE FOR MACA PLASTICS

MACA Plastics, Inc., is a woman-owned and operated plastic injection molding manufacturing firm in Adams County, Ohio. When MACA needed additional funding to expand its operations in 1995, the Ohio Valley Regional Development Commission agreed to participate in a financing package that also involved private lending and owner equity. The expansion created 18 new jobs and was so successful that the loan was paid in full by March 1997. Today MACA works closely with Cincinnati Milacron on developing new software. The company now employs 125 people and continues to grow and thrive as a major employer in the region, with $10 million in projected sales for 2001.

CONTACT:
Jeffrey Spencer
Executive Director
Ohio Valley Regional Development Commission
P.O. Box 728
Waverly, Ohio 45690-0728
(740) 947-2853
Email: jspencer@ovrdc.org

GOAL FOUR PROJECTS

PENNSYLVANIA

CREATING JOBS AND HELPING THE ENVIRONMENT: BROWNFIELDS DEVELOPMENT IN APPALACHIAN PENNSYLVANIA

Cooperative networks of government and community organizations are helping redevelop three brownfields sites in Appalachian Pennsylvania. Such areas, previously contaminated with industrial waste, often cannot be reclaimed without extensive cooperation between local, state, and federal regulators and public and private partners. The North Central Pennsylvania Regional Planning and Development Commission decided to purchase and redevelop old industrial sites, creating multitenant complexes for new companies. This move met a locally identified need for low-cost industrial space for start-up companies. As these companies grow they can relocate in the area, continuing to provide local jobs while freeing space for new entrepreneurs. The first site was acquired in 1985, and the most recent in 2001. Today the commission manages 87 acres of industrial park space, with over one million square feet of useable space occupied by 30 companies. Over 650 new jobs have been created. In addition to environmental regulatory agencies, partners have included county and municipal governments, regional planning organizations, chambers of commerce, utility companies, banks, realtors, and industrial organizations.

CONTACT:
Donald J. Masisak
Deputy Director of Economic Development
North Central Pennsylvania Regional Planning and Development Commission
651 Montmorenci Avenue
Ridgway, Pennsylvania 15853
(814) 773-3162
Email: dmasisak@ncentral.com

PUTTING IT ALL TOGETHER: SMALL BUSINESS DEVELOPMENT CENTER OUTREACH PROGRAM

For many years residents seeking to start or expand businesses in Greene County had little immediate access to local business information, counseling, or assistance in planning and financing. However, with help from the University of Pittsburgh, Greene County has implemented a key part of its strategic economic development plan. An outreach office of the University of Pittsburgh Small Business Development Center, open one day a week, was launched in late 1996. Despite its limited operation, the office assisted over 70 residents and helped six clients receive almost $1.5 million in financing in its first year. The outreach office opened on a full-time basis in 1998 and continues to provide one-on-one business management counseling, as well as informational and educational programs of interest to growing numbers of start-up or small business owners. Community involvement makes the outreach effort more successful. The Small Business Development Center is an active member of the local chamber of commerce. Its local coordinator is involved with numerous area work groups and serves as a county representative for several regional activities.

CONTACT:
Ann Dugan
Executive Director
Institute of Entrepreneurial Excellence
315 Bellefield Avenue, Room 208
Pittsburgh, Pennsylvania 15213
(412) 648-1542
Email: adugan@katz.pitt.edu

CREATING PARTNERSHIPS FOR SMALL BUSINESS GROWTH: TEAM PENNSYLVANIA ENTREPRENEURIAL ASSISTANCE NETWORKS

To build entrepreneurial strength throughout Pennsylvania's Appalachian counties and encourage development of more homegrown businesses, the state government is providing assistance through networks of private and public organizations serving entrepreneurs and small businesses in seven local development districts. The mission of these entrepreneurial assistance networks (EANs) is to create a collaborative public-private alliance to stimulate new value-added services, new market opportunities, and regional market-driven activities in order to enhance the awareness of entrepreneurship in rural areas and increase the rate of new business formation and expansion. Major EAN initiatives include Web-based client tracking systems; venture-capital training; entrepreneurial guides; placement of entrepreneurial materials in libraries; youth entrepreneurship; mentoring; linkages with medical research facilities and universities; and increased marketing of public loan programs.

CONTACT:
Neil Fowler
Acting Director
Center for Entrepreneurial Assistance
Pennsylvania Department of Community and Economic Development
Commonwealth Keystone Building, 4th Floor
Harrisburg, Pennsylvania 17120-0225
(800) 280-3801
Email: nfowler@state.pa.us

VIRGINIA

REVITALIZING A SMALL TOWN: SMALL BUSINESS INCUBATOR AT NARROWS

The town of Narrows in Giles County is fighting economic decline in a unique way. When downtown business withered due to retail competition and increased commercialization on Route 460, a major thoroughfare, town officials and community leaders decided to focus on a $1.2 million downtown redevelopment plan with the small business incubator as a primary focal point. The project was undertaken with support from the New River Valley Planning District Commission and a citizen steering committee. Today, a two-story incubator building, formerly a car dealership, offers budding businesses retail space, office suites, and light manufacturing/assembling areas. A 12,000-square-foot facility, the Giles Business Incubator had assisted in the development and growth of over 15 businesses as well as the retention and creation of over 30 jobs by the end of 2000. Narrows and Giles County recognized the importance of the incubator and began a partnership in which the county would underwrite operations for the incubator for three years, while Narrows retained ownership of the facility. Through this partnership and an aggressive marketing campaign, partially funded by a Department of Business Assistance's Small Business Incubator Program grant, the incubator is advancing from the start-up stage to the growth stage of development following the same best practices as most small businesses.

CONTACT:
Terri L. Martin
Director
Giles Business Incubator
211 Main Street, Suite 109
Narrows, Virginia 24124
(540) 726-3888
Email: gbi@gva.net

GOAL FOUR PROJECTS

DEVELOPING APPALACHIA'S FUTURE LAWYERS: THE APPALACHIAN SCHOOL OF LAW

A new law school is part of a long-term economic and civic revitalization effort in Appalachian Virginia. The long-dominant coal industry continues to decline, but community leaders in Buchanan County and surrounding counties believe the Appalachian School of Law will help build local expertise for a more diverse, complex economy. Initially housed in buildings made available by public-school consolidation, the new law school held its first graduation in 2000. Attorneys are historically a source of civic leadership, and the Appalachian School of Law seeks to attract students who will return to their communities. The school has received provisional accreditation from the American Bar Association. Beginning this autumn, the institution will become the first law school to partner with the National Bonner Leaders Program. New faculty members have been added as the school continues to grow toward its full enrollment of 350 students. Initially supported by the Appalachian Regional Commission, the law school has launched the Investment in Tomorrow development campaign, which has secured donations and pledges of over $14.4 million to fund school activities.

CONTACT:
Lucius F. Ellsworth
President
The Appalachian School of Law
P.O. Box 2825
Grundy, Virginia 24614
(540) 935-4349

SUPPORTING A SOLAR DRY KILN FACILITY IN SOUTHWESTERN VIRGINIA: APPALACHIAN SUSTAINABLE DEVELOPMENT

To encourage environmentally friendly logging and wood-products manufacturing practices, Appalachian Sustainable Development has built a solar and wood-waste dry-kiln facility to serve as a primary processing center for sustainable wood products in ten counties of southwestern Virginia and northeastern Tennessee. According to the Virginia Department of Forestry, the lack of such a facility has limited the manufacture of value-added wood products in the region. In addition to helping build the new kiln, the Appalachian Regional Commission is providing support to train loggers in sustainable harvesting methods and providing information to small manufacturers on innovative ways to market their products and expand their businesses. Appalachian Sustainable Development expects the project to help create at least four sustainable logging operations and estimates ten to 20 local wood-products manufacturers will use the newly harvested lumber.

CONTACT:
Anthony Flaccavento
Executive Director
Appalachian Sustainable Development
P.O. Box 791
Abingdon, Virginia 24212-0791
(540) 623-1121
Email: asd@naxs.net

WEST VIRGINIA

DIVERSIFYING A LOCAL ECONOMY: MINGO COUNTY WOOD PRODUCTS INDUSTRIAL PARK

Faced with a 1998 unemployment rate of over 12 percent, Mingo County officials took steps to diversify their previously coal-based local economy. Part of the county's strategy, as outlined in a new comprehensive community development plan, was to develop the Wood Products Industrial Park on 650 acres of a reclaimed former surface mine. The $28 million park project, consisting of the Wood Products Industrial Park, Appalachian Custom Dry Kilns, LLC, and Appalachian Precision Hardwood Flooring Facilities, was funded in partnership by state and federal organizations, and through conventional financing and private investment. Federal, state, and county leaders have worked cooperatively to secure funding for construction of water and wastewater treatment systems and for improvement of an access road. Thus far, the project has bolstered local economic capacity by providing a firm foothold for viable industry in the area. The project is expected to create over 150 new jobs by the time it is fully operational and to provide water and sewer services to 22 area homes.

CONTACT:
Mike Whitt
Executive Director
Mingo County Redevelopment Authority
P.O. Box 298
Williamson, West Virginia 25661
(304) 235-0042
Email: scb01005@wvnvm.wvnet.edu

FINANCING NATURAL-RESOURCE-BASED BUSINESSES: THE CONSERVATION FUND

In many distressed rural Appalachian communities the savings base cannot support new entrepreneurial ventures. With limited access to venture capital markets and few local commercial lenders, local entrepreneurs depend excessively on personal savings, retained business earnings, and support from family, friends, and business colleagues for risk capital financing. The Natural Capital Investment Fund of West Virginia (NCIF) was launched in 2000 to provide financing for high-potential, emerging natural-resource-based businesses that will advance sustainable economic growth in these rural communities. The NCIF strategy is to identify projects in which a modest amount of subordinated debt and equity can reduce risk sufficiently for commercial lenders to participate. The main targets are small businesses with strong management. The nonprofit Conservation Fund's Freshwater Institute established the NCIF in partnership with the West Virginia Small Business Development Center and the Mountain Association for Community Development, which operated a similar program in Kentucky. NCIF also works closely with the West Virginia Capital Corporation, a consortium of 56 financial institutions, and the West Virginia Jobs Investment Trust. Private foundations, the U.S.

GOAL FOUR PROJECTS

Department of Agriculture, and the Appalachian Regional Commission provided $500,000 in initial capitalization. As of June 2001, the NCIF had formed advisory and loan committees, screened over 30 business proposals, and made the first of an anticipated four initial investments.

CONTACT:
Marten R. Jenkins
Director
The Conservation Fund
Natural Capital Investment Fund
P.O. Box 1889
Shepherdstown, West Virginia 25443
(304) 876-2815
Email: m.jenkins@freshwaterinstitute.org

DEVELOPING NEW MARKETS FOR ENTREPRENEURS: CENTER FOR ECONOMIC OPTIONS

Christmas shoppers in Charleston had new choices in 2000—the products of nearly 60 small-scale entrepreneurs from throughout the West Virginia mountains. A new retail store in the Charleston Town Center indoor mall—the state's largest—was the latest effort of the Center for Economic Options, which has been helping microbusinesses find new higher-end markets since 1990. Showcase West Virginia now features the products of over 100 microbusinesses, and the center is planning two more retail locations. Over 60 percent of gross revenues go back to individual entrepreneurs. Market access has been a formidable barrier for these small-scale crafts, forest, or farm producers from isolated communities. By providing shoppers ready access to entrepreneurs, Showcase West Virginia is creating new markets that will allow self-reliance and sustainable business growth. The center partners with community representatives and organizations across the state. Other successful efforts have included the Appalachian Knitwear Project, spun off as a separate nonprofit entity, and Appalachian by Design, Inc., which began with nearly 50 home knitters producing products for the Esprit Corporation. Showcase West Virginia grew out of preparations for a subsequently cancelled West Virginia products expo. Now it helps entrepreneurs sell products while also offering them invaluable experience in cash flow, inventory control, accounting, pricing, marketing, and product development.

CONTACT:
Pam Curry
Executive Director
Center for Economic Options
214 Capital Street, Suite 200
Charleston, West Virginia 25321-0191
(304) 345-1298
Email: pcurryoptns@citynet.net

BUILDING ON ECONOMIC STRENGTHS: POLYMER ALLIANCE ZONE

Private industry and government are working together to promote and enhance the competitive advantage of the polymer industry in western West Virginia. Major companies as well as small-to-medium-sized enterprises helped design the Polymer Alliance Zone (PAZ), which provides workforce development and training, capital assistance, environmental technical assistance, and managerial assistance to small and medium-sized businesses. With a concentration of existing companies and trained workers, the area is a leading producer of polymers and polymer products. To make sure new workers are prepared for industry jobs, PAZ is collaborating with state universities and local public schools to provide pre-employment training for high school students. Working with the state, the counties, the Mid-Ohio Valley Regional Council, and West Virginia lenders, capital assistance incentives have helped create or retain over 800 jobs and over $43 million in new investments. An environmental permit handbook prepared in cooperation with the state Department of Environmental Protection provides a guide to new or revised permits. PAZ staff offers assistance to small or medium-sized companies with environmental, business, or technical problems. From multinational companies down to local recycled plastics enterprises, this unique public-private collaboration ensures continued jobs and economic growth for a major industry in three West Virginia counties.

CONTACT:
R.V. Graham
Executive Director
Polymer Alliance Zone
104 Miller Drive
Ripley, West Virginia 25271
(304) 372-1143

BRINGING EUROPE TO APPALACHIA: A REGIONAL TOURISM PROMOTION

Tourism plays an important economic role in Appalachia. In many rural communities tourism ranks among the fastest-growing sectors of the local economy. Recognizing that more repeat foreign visitors, particularly Germans, Swiss, and Austrians, were traveling to the United States, the tourism offices of West Virginia, Ohio, and Kentucky began a special coordinated effort in 1994 to promote Appalachia as an exciting and fun-filled tourism destination. Immediate steps were taken to develop relationships with airlines and to create an awareness of the Appalachian Mountain and River Region among tour operators and travel agents. At the same time, the states worked to encourage positive media coverage of the Appalachian Region in German, Swiss, and Austrian newspapers, magazines, radio, and television. Seminars were held in the three states to better prepare the tourism industries for international visitors. An Internet site, www.travelappalachia.com, provides travel information and links to the states' partners. The regional tourism project has more than met its goals. There is new international awareness of the Appalachian Region as a travel destination; more foreign visitors have come to the Region; and a local tourism industry has become increasingly attuned to the demands of expanding its own international marketing efforts.

CONTACT:
Alisa Bailey
West Virginia State Tourism Director
2101 Washington Street, East
Charleston, West Virginia 25305
(304) 558-2200
Email: abailey@callwva.com

GOAL FIVE

Health Care

Appalachian residents will have access to affordable, quality health care.

ALABAMA

PROVIDING QUALITY HEALTH CARE TO THE WORKING POOR: NORTHWEST ALABAMA COMMUNITY HEALTH CLINIC

A partnership between the University of North Alabama College of Nursing and the Florence Housing Authority is providing quality health care to the working poor in five northwestern Alabama counties. Students and faculty at the nursing college conducted a door-to-door survey in Florence public housing districts in 1996. The survey showed that many uninsured or underinsured residents had a strong need for accessible and affordable quality health care. Close to 60 percent of residents in the overwhelmingly minority areas have incomes below the national poverty level. The college, the housing authority, and other community groups founded the Northwest Alabama Community Health Association, Inc., a nonprofit corporation, to establish a nurse-managed clinic. The Florence Housing Authority provides space for the facility, which opened in 1997. The clinic offers primary health care services as well as health education in schools, churches, shelters, and recreation centers. Additional support, including a grant from the Appalachian Regional Commission, has allowed the clinic to expand its operating hours, add an additional 900 square feet to its facility, and offer dental services. Over 2,500 people are expected to receive services in 2001.

CONTACT:
Catherine Barnes
Administrative Director
Northwest Alabama Community Health Clinic
309-B Handy Homes
Florence, Alabama 35630
(256) 760-9413
Email: barnescat@aol.com

KENTUCKY

TRAINING HEALTH-CARE PROVIDERS FOR APPALACHIA: PIKEVILLE COLLEGE SCHOOL OF OSTEOPATHIC MEDICINE

While Appalachia has made substantial progress in health care in recent decades, a chronic shortage of medical professionals still exists in many rural areas. To increase the number of locally trained physicians and improve health care in Central Appalachia, the Appalachian Regional Commission has provided major support for the Pikeville College School of Osteopathic Medicine, a new medical training facility in eastern Kentucky. The Commission has helped renovate and expand school facilities and purchase additional equipment. The school, which opened in the fall of 1997, saw its first class of 53 students graduate in May 2001. Many of the new graduates expect to practice in Appalachia.

CONTACT:
John Strosnider
Dean
Pikeville College School of Osteopathic Medicine
214 Sycamore Street
Pikeville, Kentucky 41501-1194
(606) 432-9200
Email: strosnid@pc.edu

GOAL FIVE PROJECTS

STRENGTHENING THE HEALTH OF RURAL COMMUNITIES: THE UNIVERSITY OF KENTUCKY CENTER FOR EXCELLENCE IN RURAL HEALTH

When the General Assembly established the Center for Excellence in Rural Health in 1990, it purposely located the new program in Hazard. This community is in the heart of the Appalachian coalfields, one of the poorest and most rural areas of the country. As such, it suffers from all the unique and growing problems of rural health, such as chronic shortages of health professionals and inadequate access to health care information. The center designed programs to educate place-committed health professionals; encourage rural communities to take control of their health-care services; demonstrate new models of service delivery, such as cross-training health personnel; conduct applied research relating to the rural workforce; raise public health awareness; and strengthen existing services through technical assistance. Now a larger number of center graduates in medicine, nursing, physical therapy, and clinical lab sciences locate in rural areas than do graduates of any other program in the country. The center is unique in its heavy emphasis on community service and collaboration in partnership with local organizations. In 1997, the center received the Pew Award for Primary Care, and in 2000, the National Rural Health Association named it the nation's "Outstanding Rural Health Program."

CONTACT:
Loyd Kepferle
Executive Director
University of Kentucky Center for Excellence in Rural Health
100 Airport Gardens
Hazard, Kentucky 41701
(606) 439-3557

MISSISSIPPI

IMPROVING HEALTH CARE AMID SEVERE RURAL POVERTY: HICKORY FLAT CLINIC

In 1978, the only health care in Hickory Flat, a small town in Benton County, was provided by a public health nurse one day each month in one room of a dilapidated clinic building. A community committee, established that year with support from the Appalachian Regional Commission, soon organized as the Hickory Flat Clinic Association to rehabilitate, properly equip, and operate the old clinic. A full-time nurse practitioner was hired, and the renovated clinic reopened in 1979. Over the years, the clinic has become a mainstay of health care and an access point into the health care system for many patients who would otherwise avoid seeking care until serious health problems arose. An infant mortality project includes classroom education by a clinic nurse practitioner, who also serves as a health teacher at the

Hickory Flat School. The Hickory Flat Clinic averages 3,500 patient visits each year and also provides home health visits and periodic community health screenings. It is a model for other community clinics and has provided clinic experience and training for over 40 new nurse practitioners.

CONTACT:
Sue Morrisson
Director
Hickory Flat Clinic
P.O. Box 128
250 Oak Street
Hickory Flat, Mississippi 38633
(662) 333-6387
Email: suecfnp@dixie-net.com

NEW YORK

USING TECHNOLOGY TO EXPAND HEALTH CARE: TELEHOME CARE PROJECT

In southeastern New York, home health care can be costly for those with limited mobility due to chronic or terminal illnesses. Community-based nurses cannot provide the optimal number of home visits because of heavy caseloads, strained resources, and long distances. With the help of a grant from the Appalachian Regional Commission, Delaware, Otsego, and Schoharie Counties are working to remedy the problem. Using televisions and telephones, homebound patients and their nurses will soon be able to conduct telehome visits, allowing health care professionals to monitor a patient's condition via a video system. These video visits require only a fraction of the time and money needed for at-home monitoring, allowing nurses to consult with a greater number of patients over the system. The grant provides funds to train more than 70 health care professionals at four clinics and three hospital emergency rooms for this service, which continues to benefit over 100 chronically ill patients with diverse needs. This is of value not only to community health nurses but also to nurse practitioners and physician's assistants working in primary care.

CONTACT:
Kathleen Sellers
Assistant Professor of Nursing
SUNY-Utica/Rome
P.O. Box 3050
Utica, New York 13504
(315) 792-7295
Email: sellerk@sunyit.edu

GOAL FIVE PROJECTS

NORTH CAROLINA

CREATING A MODEL FOR COMMUNITY HEALTH CARE: HOT SPRINGS HEALTH PROGRAM

Providing health care through a community corporation was a new idea in the early 1970s, as was using nurse practitioners and physician's assistants as primary clinic staff. Launched by two nurses in 1971, Madison County's Hot Springs Health Program proved a trailblazer and model for health care in many other communities. A five-year grant from the Appalachian Regional Commission in 1972 helped finance the program, which was originally housed in a small, formerly abandoned physician's office. It now includes multiple medical and dental clinics and a staff of over 120, including more than 12 physicians and three nurse practitioners. It is the sole provider of primary care in the county and a principal area employer. Hot Springs has gradually evolved: in 1986, the board of directors decided the program should become self-supporting. Today it remains a community-based organization providing the first line of care to residents who once had no community health services. By early 2002, program officials expect to open a new health care center with space for five providers, as well as administrative offices for the medical, home care, and hospice services.

CONTACT:
John Graeter
Executive Director
Hot Springs Health Program
Mars Hill, North Carolina 28754
(828) 689-3471
Email: JohnG@hotspringshealth-nc.org

SHARING HEALTH INFORMATION TO IMPROVE CARE: ELECTRONIC AMBULANCE CALL REPORTING SYSTEM

In northwestern North Carolina, emergency medical service (EMS) providers are frequently unable to make informed decisions concerning a patient's prehospital care because regional health programs do not share medical records. This lack of data concerning prehospital medical treatment has led to more expensive, less efficient patient care throughout the state's Appalachian counties. Through a grant from the Appalachian Regional Commission, local health officials are developing a new telehealth database that will allow health care providers to distribute data concerning patients' medical care throughout a five-county area. Patient information will be delivered via the Internet and maintained in a private network database, saving costs, improving the delivery of services, and coordinating health programs. So far the system has accumulated over 100,000 EMS call reports. Current plans for enhancing the system include equipping EMS response vehicles with laptop computers and radio modems to aid in dispatch, vehicle locator systems, and call reporting.

CONTACT:
John C. Robertson
Special Projects Coordinator
Northwest Piedmont Council of Governments
400 West Fourth Street, Suite 400
Winston-Salem, North Carolina 27101
(336) 761-2111
Email: jrobertson@nwpcog.dst.nc.us

IMPROVING DENTAL HEALTH AMONG YOUNG CHILDREN: NORTH CAROLINA SMART SMILES PROGRAM

The chances of tooth decay have declined dramatically among 1,800 Appalachian youngsters in North Carolina as a result of an innovative program called Smart Smiles. The Appalachian Regional Commission supported a project that has recruited and trained pediatricians and their nurses, among other health professionals, to treat the teeth of pre-school children with fluoride during regular health checkups. Many young children in Appalachia suffer from severe tooth decay because their drinking water comes from wells not treated with fluoride. With a severe shortage of dentists in many rural Appalachian counties, health officials have turned to local pediatricians to offer the new fluoride treatment as part of a child's regular checkup routine.

CONTACT:
Gerry S. Cobb
Director
National Smart Start Technical Assistance Center
North Carolina Partnership for Children
1100 Wake Forest Road
Raleigh, North Carolina 27604
(919) 821-9540
Email: gscobb@smartstart-nc.org

OHIO

IMPROVING TRAINING IN HEALTH CARE: SCIOTO COUNTY JOINT VOCATIONAL SCHOOL

Discouraged by the small number of students enrolling in medical and dental training programs at the Scioto County Joint Vocational School in Appalachian Ohio, area health care advisors recommended that the school emphasize health care training and upgrade its equipment to meet training needs. The Appalachian Regional Commission has helped the school purchase new equipment including an X-ray machine, dental chairs, treatment consoles, and dental lab stations. As a result, 200 11th- and 12th-grade students and 400 adult students are benefiting annually, enrolling in a variety of courses focusing on nursing, administrative health care, dentistry, and home health care.

CONTACT:
Steve Wells
Superintendent
Scioto County Joint Vocational School
951 Vern Riffe Drive
P.O. Box 766
Lucasville, Ohio 45648
(740) 259-5522
Email: christym@scoca-k12.org

GOAL FIVE PROJECTS

PROVIDING DENTAL CARE TO LOW-INCOME RESIDENTS: SOUTHEASTERN OHIO DENTAL CLINIC

With assistance from the Appalachian Regional Commission (ARC), the Southeastern Ohio Dental Clinic in Washington County was created as a full-service clinic providing corrective and preventive dental services to low-income residents who have no other access to dental care. Serving residents of Athens, Monroe, Morgan, Noble, and Washington Counties, the clinic employs a dentist, a dental hygienist, and a dental assistant and handles more than 4,900 patient visits a year. The program has strong community support, with ARC funding matched 2-to-1 by local and state funding.

CONTACT:
David E. Brightbill
Executive Director
Community Action Program Corporation
218 Putnam Street
Marietta, Ohio 45750
(740) 373-3745
Email: Brightd@marietta.edu

TENNESSEE

MEETING A CRITICAL HEALTH NEED: COCKE COUNTY DENTAL CLINIC

In the 1996–97 school year, an oral health assessment of school children in Cocke County found tooth-decay rates 67 percent higher than the state average. Almost 30 percent of the children needed restorative or surgical treatment. These needs were not surprising given the county's poor education and poverty levels, indicators that frequently correlate with poor dental health. Clinical dental services at the Cocke County Health Department were first launched in 1991; a second dentist was added in 1997. To underscore the importance of maintaining two dentists, county officials note that the two are the only dental providers in the county for TennCare, the state-managed care program in which close to 40 percent of all county residents are enrolled. With support from the Appalachian Regional Commission, the Cocke County dental program includes education and preventive treatment and serves an average of 60 patients each month. Recently, the Cocke County Health Department received $50,000 that will be used to support new construction, underwrite remodeling of dental facilities, and contribute to the purchase of dental equipment. Officials plan to expand current building space and offer services on a full-time basis.

CONTACT:
Glenda Masters
County Director
Cocke County Health Department
430 College Street
Newport, Tennessee 37821
(423) 623-8733
Email: gmasters@mail.state.tn.us

WEST VIRGINIA

MANAGING THE HEALTH CARE NEEDS OF APPALACHIA: WEST VIRGINIA RURAL HEALTH CHRONIC DISEASE MANAGEMENT PROGRAM

Building on efforts to combat diabetes, the Chronic Disease Management Program created an effective, integrated approach to managing this and other related chronic diseases in four distressed counties. The program focused on community-based management with prescribed and supervised exercise, two hours of individualized meal planning per year by a dietician, weekly support groups coordinated by professional and lay educators, periodic cooking schools, and service coordination by registered nurses. Ebenezer Medical Outreach, Inc., in association with Marshall University, established this program to focus on health problems prevalent in the African-American community. There was candid community consultation through the Black Pastors' Association and a community diabetes forum. Partnerships were created between rural health centers and community-based groups. With approximately 20 client support groups in each county and approximately 1,300 clients assisted monthly, this program produced health improvement outcomes that were some of the most significant presented at a recent chronic disease conference. Patients involved in managing their disease received the help they needed to make and sustain lifestyle changes. The program was so successful that a how-to manual for other centers replicating the model has been prepared with support from the Appalachian Regional Commission.

CONTACT:
Richard Crespo
Executive Director
Marshall University Research Corporation
Gullickson Hall
400 Hal Greer Boulevard
Huntington, West Virginia 25755
(304) 691-1193
Email: crespo@marshall.edu

IMPROVING HEALTH THROUGH TARGETED PROGRAMS: MCDOWELL COUNTY RURAL HEALTH INITIATIVE

With support from the Appalachian Regional Commission, state and local health officials launched a coordinated effort to improve health care in McDowell County, one of the poorest counties in West Virginia. The project is composed of four health care initiatives: a free pharmaceutical program for uninsured and low-income residents; a folic-acid education pilot program for women of childbearing age; a children's health outreach program; and an emergency medical services communications enhancement program. The four initiatives have proven successful as individual components of a holistic, multifaceted approach designed to address the unique health needs of the people of McDowell County and the surrounding areas. The pharmaceutical program has provided $180,000 worth of free medicine to hundreds of county residents, and the folic-acid education pilot program has distributed 2,500 bottles of multivitamins with folic acid to women of childbearing age. The children's health outreach program has enrolled hundreds of county children, and the county has improved emergency medical services by creating new communications links between previously isolated areas and a regional medical command hospital. State officials view the health initiative as part of a larger, comprehensive capacity-building plan, designed to provide long-term solutions and build strategic partnerships that will address the county's economic and human development needs. Together the initiatives are expected to reach as many as 12,000 county residents.

CONTACT:
Sandra Pope
Director
Office of Rural Health Policy
West Virginia Department of Health and Human Resources
1411 Virginia Street, East
Charleston, West Virginia 25301
(304) 558-1327
Email: sandrapope@WVDHHR.org

Index of Projects

NORTH CAROLINA

OHIO

WEST VIRGINIA